The Modern Geometry Of The Triangle

The Modern Geometry

Of The

Triangle

Lemoine And Brocard Points – Angular And Tripolar Coordinates – Pedal And Antipedal Triangles – The Medial Triangle – Simson's Line – The Orthopole – Orthogonal Projection

By

William Gallatly, M.A.
Universities Of London And Wales

ISBN 1-933998-97-0

THE MODERN GEOMETRY OF THE TRIANGLE.

BY

WILLIAM GALLATLY, M.A.,

LATE ASSISTANT EXAMINER, UNIVERSITIES OF LONDON AND WALES.

Merchant Books

CHAPTER I.

THE LEMOINE AND BROCARD POINTS.

1. *The Lemoine Point.*—On the sides of the triangle ABC construct squares externally, and complete the diagram as given, the three outer sides of the squares meeting in A_1, B_1, C_1.

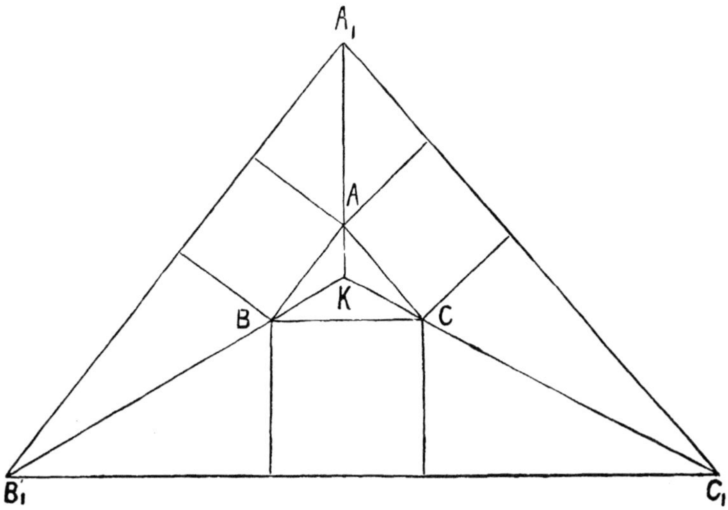

FIG. 1.

The perpendiculars from A_1 on AC, AB being b and c, the equation to AA_1 is $\beta/b = \gamma/c$.

Hence AA_1, BB_1, CC_1 meet at the point whose normal coordinates are (a, b, c), and whose barycentric coordinates are therefore (a^2, b^2, c^2).

This point is called the *Lemoine* or Grebe Point and will be denoted by K.

AK, BK, CK are called the *Symmedians* for A, B, C. The absolute values of the normal coordinates are given by $\alpha = ka$, &c., where

$$k = \frac{2\Delta}{a^2+b^2+c^2}.$$

Produce AK to meet BC in K_1; then,

$$\begin{aligned} BK_1 : CK_1 &= \triangle AKB : \triangle AKC \\ &= \text{ratio of barycentric coordinates } z \text{ and } y \\ &= c^2 : b^2; \end{aligned}$$

so that the segments BK_1, CK_1 are as the squares of adjacent sides.

2. K is the centroid of its pedal $\triangle def$.

For
$$\begin{aligned} \triangle eKf &= \tfrac{1}{2}.Ke.Kf \sin A \\ &\propto bc \sin A; \end{aligned}$$

$$\therefore \quad \triangle eKf = \triangle fKd = \triangle dKe.$$

So that K is the centroid of def.

3. If α, β, γ are the normal coordinates of any point, then $\alpha^2+\beta^2+\gamma^2$ is a minimum at K.

Since $(a^2+b^2+c^2)(\alpha^2+\beta^2+\gamma^2)$

$$\begin{aligned} &= (a\alpha+b\beta+c\gamma)^2+(b\gamma-c\beta)^2+(c\alpha-a\gamma)^2+(a\beta-b\alpha)^2 \\ &= 4\Delta^2+\dots; \end{aligned}$$

therefore $(a^2+b^2+c^2)(\alpha^2+\beta^2+\gamma^2)$ is always greater than $4\Delta^2$ except when

$$\frac{\alpha}{a} = \frac{\beta}{b} = \frac{\gamma}{c};$$

or when the point coincides with K.

In this case, therefore, $\alpha^2+\beta^2+\gamma^2$ has its minimum value, which is

$$\frac{4\Delta^2}{a^2+b^2+c^2}.$$

4. The equation to OK, joining $(\cos A, \cos B, \cos C)$ with $(\sin A, \sin B, \sin C)$ is

$$\sin(B-C)\,\alpha+\sin(C-A)\,\beta+\sin(A-B)\,\gamma = 0,$$

or

$$\frac{b^2-c^2}{a^2}\,a\alpha+\dots = 0.$$

The normal coordinates of any point on OK are given by $\alpha = m\cos(A-\theta)$, &c.

For $\Sigma \sin(B-C) \cos(A-\theta)$

$$= \cos\theta . \Sigma \cos A \sin(B-C) + \sin\theta . \Sigma \sin A \sin(B-C)$$
$$= 0.$$

5. Let figures Y and Z, directly similar, be described externally on AC, AB, so that A in Y is homologous to B in Z, and C in Y to A in Z.

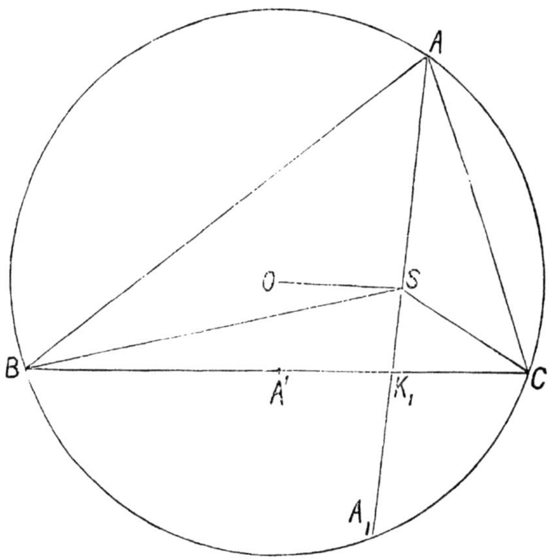

FIG. 2.

Then, if S be the double point of Y and Z, the triangles SAC, SBA are similar, having

$$\text{angle } SBA = SAC;\ SAB = SCA;\ ASB = ASC,$$

so that S is the focus of a parabola touching AB, AC at B, C.

Let p_2, p_3 be the perpendiculars from S on AC, AB.

Then, from similar triangles SBA, SAC

$$p_2 : p_3 = AC : AB,$$

so that S lies on the A-symmedian.

Again, angle $BSK_1 = SAB + SBA$

$$= SAB + SAC$$
$$= A;$$

$$\therefore\quad BSC = 2BSK_1 = 2A = BOC;$$

so that S lies on the circle BOC.

Also, $OSK_1 = OSB + A = OCB + A$ (in circle $BOSC$)

$= 90°$;

$\therefore \quad SA = SA_1$.

Hence the double point S of the two similar figures on AB, AC may be found by drawing the chord AA_1 through K and bisecting it at S.

6. Let $T_1T_2T_3$ be the Tangent Triangle, formed by drawing tangents to the circumcircle at A, B, C.

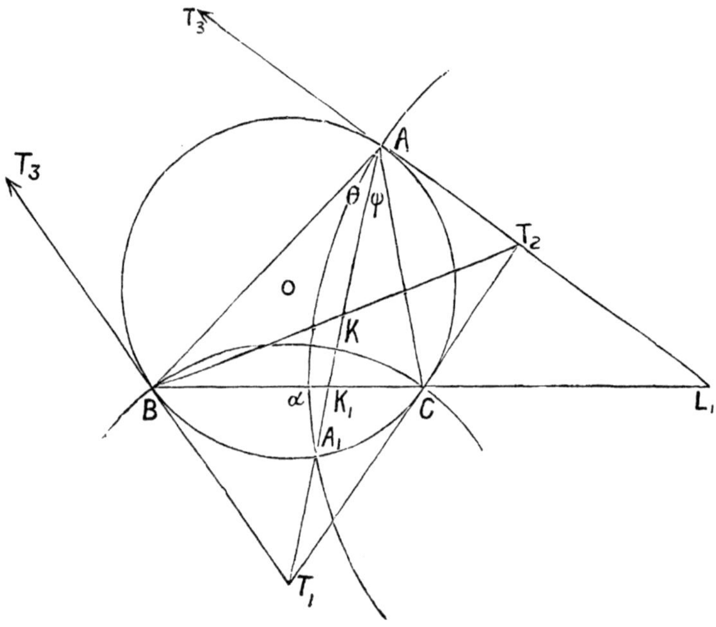

FIG. 3.

Then AT_1, BT_2, CT_3 are concurrent at K.

For, if q, r be the perpendiculars from T_1 on AC, AB,

$$\frac{q}{r} = \frac{T_1C \sin ACT_2}{T_1B \sin ABT_3} = \frac{\sin B}{\sin C} = \frac{b}{c}.$$

Hence AT_1 passes through K; so also do BT_2, CT_3.

Since ABC is the inscribed circle of $T_1T_2T_3$, the point K is the Gergonne Point of this triangle.

7. To prove that AK bisects all chords of the $\triangle ABC$, which are parallel to the tangent at A, or perpendicular to OA, or parallel to the side H_2H_3 of the orthic $\triangle H_1H_2H_3$.

Let $BAK = \theta$, $CAK = \phi$.

Then $$\frac{\sin\theta}{\sin\phi} = \frac{\gamma}{\beta} = \frac{c}{b} \text{ : a \textit{most important} relation ;}$$
$$= \frac{\sin C}{\sin B} = \frac{\sin BAT_2}{\sin CAT_2}.$$

Therefore AK and T_2AT_3 are harmonic conjugates with respect to AB and AC.

It follows that AK bisects all chords which are parallel to T_2T_3 or H_2H_3, or are perpendicular to OA.

8. *The Apollonian Circles.*—Let the several pairs of bisectors of the angles A, B, C meet BC in α, α' ; CA in β, β' ; AB in γ, γ'. (See Fig. 3.)

The circles described on $\alpha\alpha'$, $\beta\beta'$, $\gamma\gamma'$ as diameters are called the Apollonian Circles.

Let the tangent at A to the circle ABC meet BC in L_1.

Then $$\text{angle } L_1\alpha A = \alpha BA + BA\alpha = B + \tfrac{1}{2}A,$$
$$L_1A\alpha = L_1AC + CA\alpha = B + \tfrac{1}{2}A \text{ ;}$$
so that $$L_1\alpha = L_1A.$$
And, since $$\alpha A\alpha' = 90^\circ,$$
$$\therefore \quad L_1A = L_1\alpha' \text{ ;}$$
so that L_1 is the centre of the Apollonian circle $(\alpha\alpha')$, passing through A and orthogonal to the circle ABC.

Since AL_1 is a tangent, the polar of L_1 passes through A ; and, since (BK_1CL_1) is harmonic (7) *, the polar of L_1 passes through K_1.

It follows that $AKK_1A_1T_1$ is the polar of L_1 ; so that L_1A_1 is the other tangent from L_1.

Hence the common chords AA_1, BB_1, CC_1 of the circle ABC and the Apollonian Circles intersect at K, which is therefore equipotential for the four circles.

Note that OL_1 bisects AKK_1A_1 at right angles, and therefore passes through the point S of Section (5).

The Lemoine Line.—Since the polars of L_1, L_2, L_3 pass through K, the points L_1, L_2, L_3 are collinear, lying on the polar of K. This polar is found by the ordinary method to be
$$\alpha/a + \beta/b + \gamma/c = 0,$$
and is called the Lemoine Line or Axis.

* The bracketed numbers refer to sections.

9. The Lemoine Point is the point of intersection of lines joining the mid-points of the sides of ABC to the mid-points of corresponding perpendiculars.

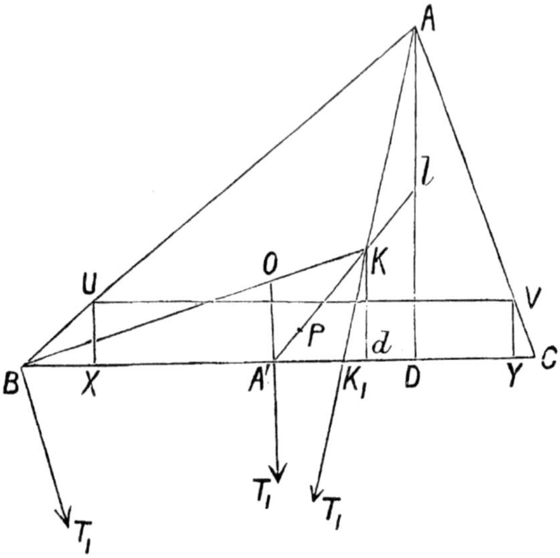

Fig. 4.

Let A', B', C' be the mid-points of BC, CA, AB, and let $A'K$ meet AD in l.

Then, since BK and the tangent BT_1 at B form a harmonic pencil with BA, BC, therefore the range (AKK_1T_1) is harmonic; therefore the pencil $A'(AlD\infty)$ is harmonic, so that AD is bisected at l.

10. A rectangle $XYVU$ being inscribed in ABC with the side XY on BC, to find the locus of its centre P.

The diagram shows that the mid-point of UV lies on $A'A$, and that P lies on $A'l$.

The Lemoine Point K, common to $A'l$, $B'm$, $C'n$ is the common centre of three inscribed rectangles, standing on BC, CA, AB respectively.

11. To determine the "direction angles," θ_1, θ_2, θ_3, which OK makes with the sides of ABC.

From the diagram,

$$OK \cos \theta_1 = A'd = A'D \frac{Kd}{lD}.$$

Now $$A'D = \frac{b^2-c^2}{2a}, \quad Kd = \frac{2\Delta}{a^2+b^2+c^2}\,a,$$

$$lD = \tfrac{1}{2}h_1 = \frac{\Delta}{a};$$

so that $$\cos\theta_1 \propto a\,(b^2-c^2);$$

$$\therefore \quad \cos\theta_1 : \cos\theta_2 : \cos\theta_3 = a\,(b^2-c^2) : b\,(c^2-a^2) : c\,(a^2-b^2).$$

12. *The Harmonic Quadrilateral.*—The angles of the harmonic pencil at A are seen to be $B\phi\theta$ or $C\theta\phi$. (See Fig. 3.)

The same angles are found, in the same order, at B and C and A_1.

Hence the pencils at B, C, A, and A_1 are harmonic; ABA_1C being called, on this account, a Harmonic Quadrilateral.

In the triangle ABA_1 the tangent at B is harmonically conjugate to BK_1, so that BK_1 is the B-symmedian for this triangle.

Similarly, A_1K_1 is the A_1-symmedian in BA_1C_1; and CK_1 the C-symmedian in ACA_1.

To prove that rectangle $AB.A_1C$ = rectangle $AC.A_1B$,

$$\frac{AB}{AC} = \frac{\sin\theta}{\sin\phi} \qquad (7)$$

$$= \frac{\sin BCA_1}{\sin CBA_1} = \frac{A_1B}{A_1C}.$$

$$\therefore \quad AB.A_1C = AC.A_1B.$$

If x, y, z, u are the perpendiculars from K_1 on AB, AC, A_1B, A_1C, then

$$\frac{x}{AB} = \frac{y}{AC} = \frac{z}{A_1B} = \frac{u}{A_1C}.$$

For, since AK_1 is the A-symmedian of ABC,

$$\therefore \quad \frac{x}{AB} = \frac{y}{AC}.$$

And, since BK_1 is the B-symmedian of $\triangle ABA_1$,

$$\frac{x}{AB} = \frac{z}{A_1B}.$$

And, since A_1K_1 is the A_1-symmedian of A_1BC,

$$\therefore \quad \frac{z}{A_1B} = \frac{u}{A_1C}.$$

13. *The Brocard Points.*—On the sides BC, CA, AB let triads of circles be described whose *external segments* contain the angles

$$(a)\ A,\ B,\ C;$$
$$(b)\ C,\ A,\ B;$$
$$(c)\ B,\ C,\ A;$$

the cyclic order (ABC) being preserved.

The first triad of circles intersect at the orthocentre H.

Let the second triad intersect at Ω, and the third at Ω'.

Ω and Ω' are the *Brocard Points* of ABC.

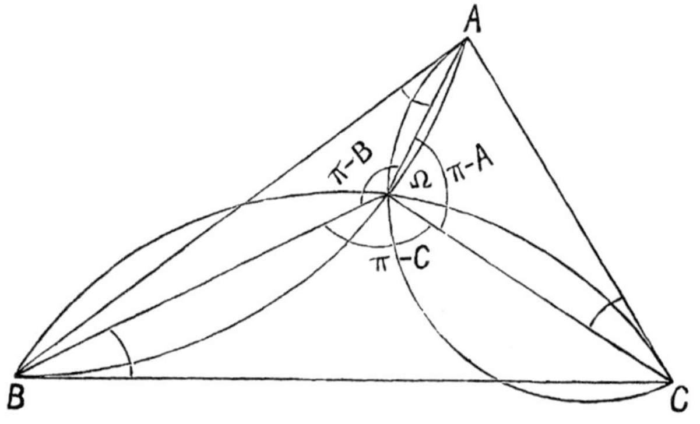

FIG. 5.

Since the external segment of $B\Omega C$ contains the angle C, this circle touches CA at C.

Similarly, $C\Omega A$ touches AB at A, and $A\Omega B$ touches BC at B.

(Memorize the order of angles for Ω by the word "CAB.")

In like manner, $B\Omega'C$ touches AB at B, $C\Omega'A$ touches BC at C, $A\Omega'B$ touches CA at A.

Again, since $A\Omega B$ touches BC at B, the angle

$$\Omega BC = \Omega AB.$$

So $$\Omega CA = \Omega BC.$$

Similarly, $$\Omega' AC = \Omega' CB = \Omega' BA.$$

Denote each of the equal angles ΩBC, ΩCA, ΩAB by ω, and each of the equal angles $\Omega' CB$, &c., by ω'.

14. To determine ω and ω'.

In $\triangle A\Omega B$, $$\Omega B = c\,\frac{\sin\omega}{\sin B} = 2R\sin\omega\,\frac{c}{b}.$$

So $$\Omega A = 2R \sin \omega \, \frac{b}{a};$$

$$\Omega C = 2R \sin \omega \, \frac{a}{c}.$$

Hence $$\frac{\sin(A-\omega)}{\sin \omega} = \frac{\Omega C}{\Omega A} = \frac{a^2}{bc}.$$

$$\therefore \quad \cot \omega = \frac{a^2+b^2+c^2}{4\Delta}$$

$$= \cot A + \cot B + \cot C$$

$$= \frac{1+\cos A \cos B \cos C}{\sin A \sin B \sin C}.$$

The same expressions are found for $\cot \omega'$;

$$\therefore \quad \omega' = \omega.$$

The angle ω, which is equal to each of the six angles ΩAB, ΩBC, ΩCA, $\Omega' BA$, $\Omega' CB$, $\Omega' AC$, is called the *Brocard Angle* of ABC.

15. To determine the normal and barycentric coordinates of Ω and Ω'.

From the diagram,

$$\alpha = \Omega B \sin \omega = 2R \sin^2 \omega \, \frac{c}{b}.$$

So $$\beta = 2R \sin^2 \omega \frac{a}{c}; \quad \gamma = 2R \sin^2 \omega \, \frac{b}{a}.$$

And for Ω',

$$\alpha' = 2R \sin^2 \omega \, \frac{b}{c}; \quad \beta' = 2R \sin^2 \omega \frac{c}{a}; \quad \gamma' = 2R \sin^2 \omega \frac{a}{b}.$$

The barycentric coordinates of Ω are given by

$$x : y : z = \frac{1}{b^2} : \frac{1}{c^2} : \frac{1}{a^2};$$

and for Ω' $$x' : y' : z' = \frac{1}{c^2} : \frac{1}{a^2} : \frac{1}{b^2}.$$

16. The Brocard angle is never greater than 30°.

For $$\cot \omega = \cot A + \cot B + \cot C,$$

and $$\cot B \cot C + \ldots = 1;$$

$$\therefore \quad \cot^2 \omega = \cot^2 A + \ldots + 2;$$

$$\therefore \quad (\cot B - \cot C)^2 + \ldots = 2(\cot^2 A + \ldots) - 2$$
$$= 2(\cot^2 \omega - 3).$$

Hence $\cot \omega$ is never less than $\sqrt{3}$, and therefore ω is never greater than 30°.

17. Some useful formulæ:—

(i)
$$\operatorname{cosec}^2 \omega = 1 + \cot^2 \omega = 1 + \frac{(a^2+b^2+c^2)^2}{16\Delta^2}$$
$$= \frac{b^2c^2 + c^2a^2 + a^2b^2}{4\Delta^2},$$

and
$$1 - 4\sin^2 \omega = \frac{a^4+b^4+c^4-b^2c^2-c^2a^2-a^2b^2}{b^2c^2+c^2a^2+a^2b^2}.$$

(ii) Since
$$\frac{\sin(A-\omega)}{\sin \omega} = \frac{a^2}{bc},$$
$$\therefore \quad \sin(A-\omega) : \sin(B-\omega) : \sin(C-\omega) = a^3 : b^3 : c^3.$$

(iii)
$$\frac{\sin(A+\omega)}{\sin \omega} = \sin A(\cot \omega + \cot A)$$
$$= \frac{b^2+c^2}{bc}.$$

Note that, when $b = c$, $\sin(A+\omega) = 2\sin\omega$.

(iv)
$$\frac{\cos(A+\omega)}{\sin A \sin \omega} = \cot A \cot \omega - 1$$
$$= \frac{(-a^2+b^2+c^2)(a^2+b^2+c^2)}{16\Delta^2} - 1.$$
$$\therefore \quad \cos(A+\omega) = \frac{\sin A \sin \omega}{8\Delta^2}\{b^4+c^4-a^2b^2-a^2c^2\}$$
$$\propto \sin(A-B)\sin B + \sin(A-C)\sin C.$$

CHAPTER II.

ANGULAR AND TRIPOLAR COORDINATES.

18. Take any point P *within the triangle* ABC.
Let angle $BPC = X$, $CPA = Y$, $APB = Z$.
The angles X, Y, Z are called the Angular Coordinates of P.
To determine the normal and barycentric coordinates of P.
Let AP, BP, CP intersect the circle ABC in L, M, N.
Since $BPC = X$, we have

$$MPC = 180-X, \quad PMC = A, \quad PCM = X-A.$$

$$\therefore \quad PC = PM\frac{\sin A}{\sin(X-A)};$$

$$\therefore \quad a\alpha = 2\triangle BPC = BP.PC.\sin X$$
$$= BP.PM\frac{\sin X \sin A}{\sin(X-A)}$$
$$= \Pi\frac{\sin X \sin A}{\sin(X-A)},$$

where Π = power of P for circle ABC;

or
$$2R.\alpha = \Pi\frac{\sin X}{\sin(X-A)}.$$

19. Let λ, μ, ν be the angles of LMN.
It is required to determine the power and the coordinates of P in terms of λ, μ, ν.

We have $X = BPC = BMC + MCN = A + \lambda$;

and now
$$a\alpha = \Pi\frac{\sin(A+\lambda)\sin A}{\sin\lambda},$$

and
$$2R\alpha = \Pi\frac{\sin(A+\lambda)}{\sin\lambda}.$$

$$\therefore \quad 2\Delta/\Pi = \Sigma\sin^2 A\cot\lambda + \tfrac{1}{2}\Sigma\sin 2A;$$

$$\therefore \quad 8R^2\Delta/\Pi \text{ or } 2R\,abc/\Pi = a^2\cot\lambda + b^2\cot\mu + c^3\cot\nu + 4\Delta,$$

an expression which will frequently present itself in this work. Denote it by Q.*

$$\therefore \quad \Pi = \frac{2R\,abc}{Q} \quad \text{or} \quad \frac{\Pi}{R^2} = \frac{8\Delta}{Q},$$

and
$$a = \frac{abc}{Q}\,\frac{\sin(A+\lambda)}{\sin\lambda}.$$

20. *Inversion.*—To invert a triangle ABC into a triangle with given angles λ, μ, ν.

Determine a point P lying *within the circle* ABC whose angular coordinates are

$$X = A+\lambda, \quad Y = B+\mu, \quad Z = C+\nu.$$

Produce AP, BP, CP to meet the circle again in L, M, N.

Then $\qquad L = \lambda, \quad M = \mu, \quad N = \nu. \quad (19)$

Also $\qquad PA.PL = PB.PM = PC.PN = \Pi.$

Hence ABC is inverted into the $\triangle LMN$, with given angles λ, μ, ν; the pole being P, and $\sqrt{\Pi}$ being the radius of inversion.

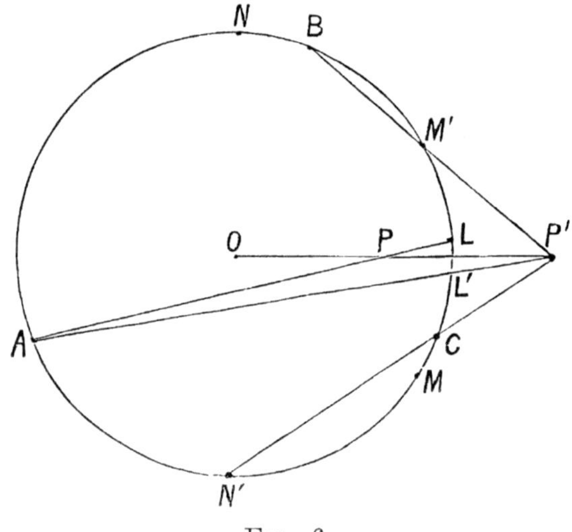

FIG. 6.

21. There is also a *second* pole of inversion.

The point P', inverse to P, lies on OP, and is such that

$$OP.OP' = R^2.$$

---

* The expression $a^2 \cot\lambda + \ldots$ was first used, I believe, by Dr. J. Schick, Professor in the University of Munich. The relations he deals with are different from those given in this work.

Let $P'A$, $P'B$, $P'C$ meet the circle in L', M', N'.

Then, since $$OL^2 = OP.OP',$$

the angle $$OP'L = OLP \text{ or } OLA = OAP$$
$$= OP'A \text{ or } OP'L' \quad (OA^2 = OP.OP'),$$

so that L' is the image of L in OPP'.

Similarly, M', N' are the images of M and N, and thus the triangles LMN, $L'M'N'$ are in all respects equal, though inverse in position.

We then have
$$L' = L = \lambda, \quad M' = M = \mu, \quad N' = N = \nu;$$

and, since $P'A.P'L' = P'B.P'M' = P'C.P'N' = \Pi'$,

the triangle ABC is inverted into $L'M'N'$, with angles λ, μ, ν; the pole being P', and the radius of inversion $\sqrt{\Pi'}$.

22. To determine the angular coordinates X', Y', Z' and the power Π' of P' in terms of $\lambda\mu\nu$.

In all cases X, Y, Z are to be measured round *in the same direction*, so that $X+Y+Z = 360°$ always.

From the diagram,
$$Y' = AP'C = CL'A - L'CP'$$
$$= CL'A - L'AN' \quad \text{(cyclic quadrilateral } CL'AN'\text{)}$$
$$= B - \mu.$$

So $$Z' = C - \nu;$$

$$\therefore \quad X' = 360° - Y' - Z' = 360° - (\lambda - A).$$

And $$aa' = \Pi' \frac{\sin X' \sin A}{\sin (X' - A)}$$
$$= \Pi' \frac{\sin (A - \lambda) \sin A}{\sin \lambda}.$$

Hence, as in (19), $$\Pi' = \frac{2R\,abc}{Q'},$$

where $$Q' = a^2 \cot \lambda + b^2 \cot \mu + c^2 \cot \nu - 4\Delta$$

and $$a' = \frac{abc}{Q'} \frac{\sin (A - \lambda)}{\sin \lambda}.$$

23. *Orthologic Triangles.*—Take any point P, and draw any straight lines YZ, ZX, XY perpendicular to PA, PB, PC; so that the perpendiculars from A, B, C on the sides of XYZ are concurrent (at P).

Then shall the perpendiculars from X, Y, Z be concurrent.

We have
$$PY^2-PZ^2 = AY^2-AZ^2;$$
$$PZ^2-PX^2 = BZ^2-BX^2;$$
$$PX^2-PY^2 = CX^2-CY^2;$$
$$\therefore \quad (BX^2-CX^2)+(CY^2-AY^2)+(AZ^2-BZ^2) = 0.$$

Hence the perpendiculars from X, Y, Z on BC, CA, AB are concurrent, say at Q.

Triangles ABC, XYZ, which are thus related, are said to be mutually Orthologic.

To determine the trigonometrical relation between P and Q.

Since PB, PC are perpendicular to XZ, XY,
$$\therefore \quad \text{angle } BPC = \pi - X;$$
so that, if x, y, z are the barycentric coordinates of P in ABC,
$$x \propto \frac{\sin(\pi-X)\sin A}{\sin(\pi-X-A)} \propto \frac{\sin X \sin A}{\sin(X+A)}.$$

Again for the triangle XYZ,
$$YQZ = \pi - A;$$
so that, if x', y', z' are the barycentric coordinates of Q in XYZ,
$$x' \propto \frac{\sin(\pi-A)\sin X}{\sin(\pi-A-X)} \propto \frac{\sin X \sin A}{\sin(X+A)}.$$

Hence the barycentric coordinates of P in ABC are proportional to the barycentric coordinates of Q in XYZ; the ratio evidently being that of the areas of the triangles.

24. A Harmonic Quadrilateral such as ABA_1C can be inverted into a square.

Let the circle described with centre T_1 and radius T_1B or T_1C cut the Apollonian Circle L_1 in E and E'.

Then, since the tangents OA, OB, OC, OA_1 to the two circles are equal, O lies on their common chord EE', and
$$OE \, . \, OE' = R^2;$$
so that E, E' are inverse points for the circle ABC.

Now the angle $BT_1C = \pi - 2A$;
$$\therefore \quad BEC = \pi - \tfrac{1}{2}(\pi - 2A) = A + \tfrac{1}{2}\pi.$$

Hence, if ABC be inverted from E as pole into LMN,

the angle L is a right angle ($\lambda = \frac{1}{2}\pi$). (19)

Again, in the triangle ABA_1, the circle AEA_1 having centre L_1 and radii L_1A and L_1A_1 touching circumcircle ABA_1, corresponds to the circle BEC with centre T_1 in the triangle ABC.

And, since BK_1C is the B-symmedian of the triangle ABA_1, it follows that the circle BEC with centre T_1 is the Apollonian circle of ABA_1 for the side AA_1.

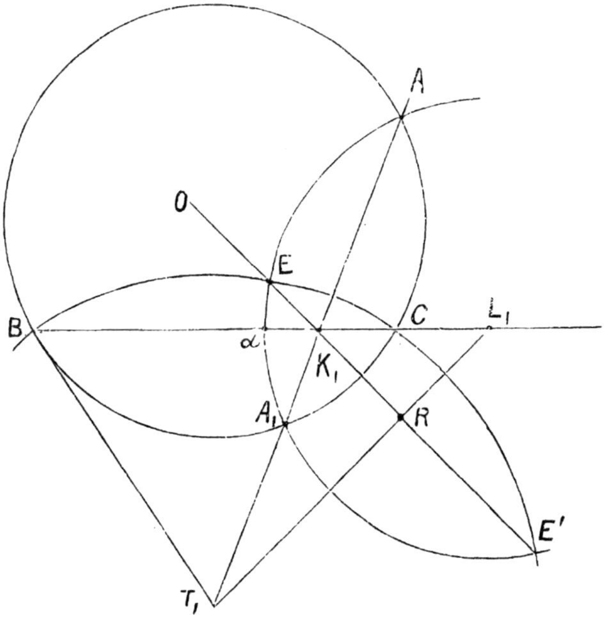

FIG. 7.

Hence E, where these two circles intersect, holds the same position in ABA_1 as it holds in ABC; so that, when ABA_1 is projected into LML', with E as pole of inversion, the angle M is a right angle, and so on.

Hence, $LML'N$ is a rectangle.

And since, in the inverted figure, the pencils at the angular points remain harmonic, the rectangle is a square.

Another square is obtained by inverting from E'.

25. *Tripolar Coordinates.*—The tripolar coordinates of a point are the distances, or the ratios of the distances, of the point from A, B, C.

To determine the point whose tripolar coordinates are l, m, n.

Divide the sides BC, CA, AB at P, Q, R,

so that
$$BP : PC = m : n,$$
$$CQ : QA = n : l,$$
$$AR : RB = l : m.$$

Then AP, BQ, CR are concurrent at a point L whose barycentric coordinates are $1/l$, $1/m$, $1/n$.

Produce QR to meet BC in P'.

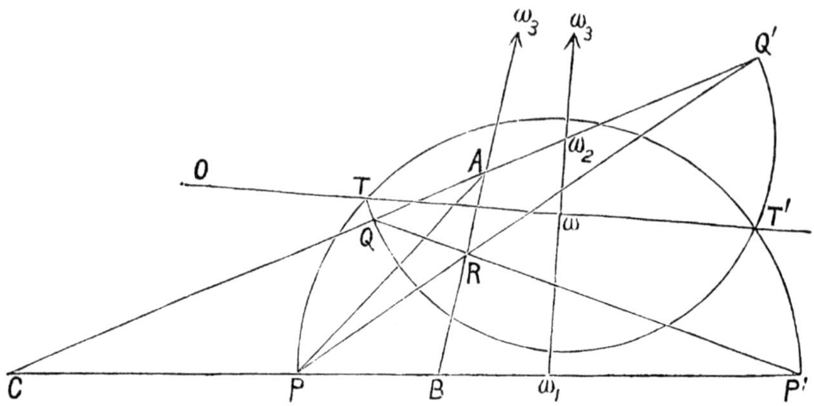

FIG. 8.

Then, since AP is $my-nz=0$ in barycentric coordinates, and the range $CPBP'$ is harmonic, it follows that AP' is $my+nz=0$; so that P', Q', R' lie on the line

$$lx+my+nz=0.$$

Bisect PP' in ω_1; then

$$\omega_1 P^2 = \omega_1 P'^2 = \omega_1 B \, . \, \omega_1 C,$$

since $(CPBP')$ is harmonic.

Hence the circle ABC cuts the circles (PP'), (QQ'), (RR') orthogonally.

Again
$$\frac{m^2}{n^2} = \frac{BP^2}{CP^2} = \frac{BP'^2}{CP'^2} = \frac{BP'^2 - BP^2}{CP'^2 - CP^2}$$

$$= \frac{B\omega_1}{C\omega_1},$$

so that ω_1, ω_2, ω_3 lie on the line

$$l^2x + m^2y + n^2z = 0.$$

Since $(CPBP')$ is harmonic, for every point T on the circle (PP') we have $BT : CT : = BP : CP = m : n$.

So, for every point T on (QQ'),

$$CT : AT = n : l.$$

Therefore at the points T, T', where these circles intersect,

$$AT : BT : CT = l : m : n = AT' : BT' : CT'.$$

The symmetry of the result shows that the third circle (RR') also passes through T and T'.

Since these circles (PP'), (QQ'), (RR') cut the circle ABC orthogonally, and their centres ω_1, ω_2, ω_3 are collinear, it follows that their common Radical Axis TT' passes through O; and therefore OTT' is perpendicular to $\omega_1\omega_2\omega_3$,

$$OT.OT' = R^2,$$

and every circle through T and T' cuts the circle ABC orthogonally.

Let the circle (TT'), whose centre is ω, cut the circle ABC in M, M', so that the angle $\omega MO = 90°$, and let MM' cut $O\omega$ in J.

Then $$OJ.O\omega = OM^2 = R^2,$$

so that J is the pole of $\omega_1\omega_2\omega_3$ for the circle ABC.

Also $$\omega M^2 = \omega J.\omega O,$$

so that $\omega_1\omega_2\omega_3$ is the Radical Axis of circles ABC and (OJ).

Note also that $OTJT'$ is harmonic.

26. As an illustration, consider the *Isodynamic points* U *and* U'.

These are the positions of T, T' when L coincides with the in-centre I.

Then $$BP : CP = c : b,$$

and $$l : m : n = 1/a : 1/b : 1/c;$$

so that (UA, UB, UC) and $(U'A, U'B, U'C)$ are inversely as (a, b, c).

The line $P'Q'R'$ is then

$$\frac{x}{a} + \ldots = 0 \quad \text{or} \quad \alpha+\beta+\gamma = 0,$$

the Radical Axis of ABC and $I_1I_2I_3$.

The line of centres $\omega_1\omega_2\omega_3$ is then

$$\frac{x}{a^2} + \ldots = 0 \quad \text{or} \quad \frac{\alpha}{a} + \frac{\beta}{b} + \frac{\gamma}{c} = 0,$$

which is the Lemoine Line; and J, the pole of this line, coincides with K.

The points U, U' therefore *lie on* OK, and are harmonic conjugates with O and K.

27. To determine the tripolar equation to a circumdiameter, which makes angles θ_1, θ_2, θ_3 with the sides.

Let P be any point on the line, xyz the projections of OP on the sides.

Then if r_1, r_2, r_3 be the tripolar coordinates of P,

$$r_2^2 - r_3^2 = a.2x = a.2.OP \cos\theta.$$

And $$(r_2^2-r_3^2)r_1^2+(r_3^2-r_1^2)r_2^2+(r_1^2-r_2^2)r_3^2=0;$$

$$\therefore \quad a\cos\theta_1.r_1^2+b\cos\theta_2.r_2^2+c\cos\theta_3.r_3^2=0,$$

and $$a\cos\theta_1+\ldots=0.$$

So that an equation of the form

$$lr_1^2+mr_2^2+nr_3^2=0,$$

where $$l+m+n=0,$$

represents a circumdiameter.

Note that, for every point P or (r_1, r_2, r_3) on the line, the ratios $r_2^2-r_3^2 : r_3^2-r_1^2 : r_1^2-r_2^2$ are constant; being, in fact, equal to the ratios $a\cos\theta_1 : b\cos\theta_2 : c\cos\theta_3$.

28. The angles θ_1, θ_2, θ_3, made by a straight line OT with the sides of ABC, will henceforth be called the Direction Angles of the line.

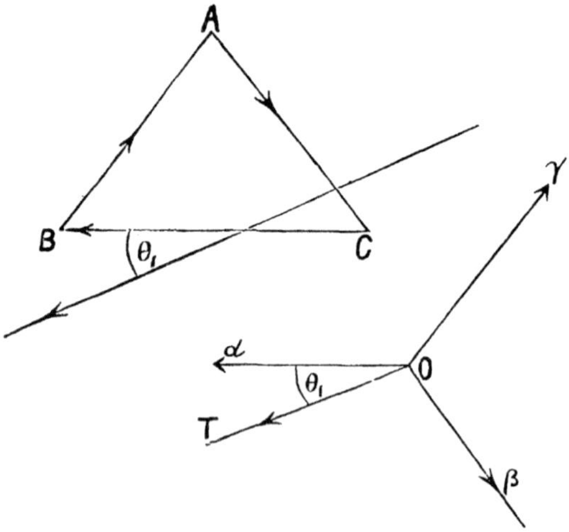

FIG. 9.

It is *very important* to note that these direction angles must be measured in the same sense, from the given line as axis.

Draw $O\alpha$, $O\beta$, $O\gamma$ parallel to, and in the *same sense* with, BC, CA, AB or CB, BA, AC.

Then, if $TO\alpha$ is taken as θ_1 (or the supplement of $TO\alpha$ may be taken as θ_1)

$$TO\gamma = TO\alpha + \alpha O\gamma = \theta_3,$$
$$TO\beta = TO\gamma + \gamma O\beta = \theta_2,$$
$$\alpha O\gamma = \pi - B, \quad \gamma O\beta = \pi - A.$$

Then, by projecting the sides on to the line and perpendicular to it, we obtain

$$a\cos\theta_1 + b\cos\theta_2 + c\cos\theta_3 = 0,$$
$$a\sin\theta_1 + b\sin\theta_2 + c\sin\theta_3 = 0.$$

29. The tripolar equations to OI, OGH, OK, $O\Omega$ should be noted.

(i) The projection of OI on $BC = \frac{1}{2}(b-c)$;

$$\therefore \quad \cos\theta_1 \propto (b-c);$$

and the equation to OI is

$$a(b-c)r_1^2 + \ldots = 0.$$

(ii) The projection of OH on $BC = \dfrac{b^2-c^2}{2a}$;

and the equation to OH is

$$(b^2-c^2)r_1^2 + \ldots = 0.$$

(iii) The projection of OK on BC

$$\propto a(b^2-c^2); \qquad (11)$$

and the equation to OK is

$$a^2(b^2-c^2)r_1^2 + \ldots = 0.$$

(iv) Noting that $B\Omega \propto \dfrac{c}{b}$, $C\Omega \propto \dfrac{a}{c}$;

and that $r_2^2 - r_3^2 \propto a\cos\theta_1$,

we find for the equation of $O\Omega$

$$a^2(c^4 - a^2b^2)r_1^2 + b^2(a^4 - b^2c^2)r_2^2 + c^2(b^4 - c^2a^2)r_3^2 = 0.$$

Observe—

That the point whose tripolar coordinates are $(s-a, s-b, s-c)$ lies on OI;

That $(a, b, c,)$ and $(\cos A, \cos B, \cos C)$ lie on OH;

That $(1/a, 1/b, 1/c)$ and $(\cot A, \cot B, \cot C)$ lie on OK.

30. To find the equation to a straight line with direction angles θ_1, θ_2, θ_3, and at a distance d from O.

Transferring to Cartesian coordinates, we see that the equation differs only by a constant from that of the parallel diameter.

It must therefore be of the form

$$a\cos\theta_1 . r_1^2 + \ldots = k.$$

Let A', B', C' be the mid-points of the sides, and let $A'O$ meet the line in A''.

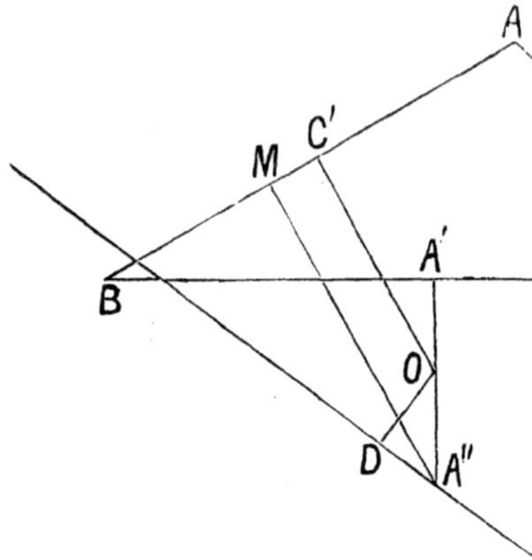

FIG. 10.

Then if (ρ_1, ρ_2, ρ_3) be the coordinates of A'',

$$\begin{aligned} k &= a\cos\theta_1 . \rho_1^2 + b\cos\theta_2 . \rho_2^2 + c\cos\theta_3 . \rho_3^2 \quad [\rho_2 = \rho_3] \\ &= a\cos\theta_1 (\rho_1^2 - \rho_2^2) \\ &= a\cos\theta_1 (AM^2 - MB^2) \\ &= a\cos\theta_1 . 2c . C'M \\ &= a\frac{OD}{OA''} 2c . OA'' \sin B \\ &= d . 2ac \sin B \\ &= 4d\Delta. \end{aligned}$$

Hence the required equation is

$$a\cos\theta_1 . r_1^2 + \ldots = 4d\Delta.$$

31. When $(l+m+n)$ is *not zero.*

To prove that, if Q be the mean centre of masses l, m, n at A, B, C, or if (l, m, n) are the barycentric coordinates of Q, then, for any point P whose tripolar coordinates are (r_1, r_2, r_3),

$$lr_1^2 + mr_2^2 + nr_3^2 = l . AQ^2 + m . BQ^2 + n . CQ^2 + (l+m+n)PQ^2.$$

Take any rectangular axes at Q, and let $(a_1 a_2)$, $(b_1 b_2)$, $(c_1 c_2)$, (xy) be the Cartesian coordinates of A, B, C, P.

Then $$\begin{aligned} lr_1^2 &= l(a_1 - x)^2 + l(a_2 - y)^2 \\ &= l . AQ^2 + l . PQ^2 - 2x . la_1 - 2y . la_2. \end{aligned}$$

But, since Q is the mean centre for masses l, m, n,

$$\therefore \quad la_1+mb_1+nc_1 = 0, \quad la_2+mb_2+nc_2 = 0;$$

$$\therefore \quad lr_1^2+mr_2^2+nr_3^2 = l.AQ^2+m.BQ^2+n.CQ^2+(l+m+n)\,PQ^2.$$

This, being true for *any* point P, is true for 0.

$$\therefore \quad l.R^2+m.R^2+n.R^2 = \Sigma\, l.AQ^2+(l+m+n)\;OQ^2;$$

$$\therefore \quad lr_1^2+mr_2^2+nr_3^2 = (l+m+n)(QP^2-QO^2+R^2).$$

32. *Particular Cases.*—

(i) If P so move that $\Sigma\, l.r_1^2 = \text{constant} = k^2$, then QP is constant; so that P describes a circle with centre Q and radius ρ, given by

$$\rho^2 = QP^2 = k^2/(l+m+n)+QO^2-R^2.$$

(ii) If $lr_1^2+mr_2^2+nr_3^2 = 0$;
then, as Mr. R. F. Davis has pointed out,

$$\rho^2 = QP^2 = QO^2-R^2;$$

so that the circle cuts the circle ABC *orthogonally*.

(iii) If the circle passes through O,
then $QP = QO$; and the equation becomes

$$lr_1^2+mr_2^2+nr_3^2 = (l+m+n)\;R^2.$$

(iv) When P coincides with Q, $QP = 0$,
and $l.QA^2+m.QB^2+n.QC^2 = (l+m+n)\times$power of Q,

so that the power of $Q = \dfrac{l.QA^2+m.QB^2+n.QC^2}{l+m+n}$.

Examples.—

(A) The Circumcircle:

Here $\qquad QP = R, \quad QO = 0, \quad l \propto \sin 2A;$

$$\therefore \quad \sin 2A.r_1^2+\sin 2B.r_2^2+\sin 2C.r_3^2 = 4\Delta.$$

(B) The inscribed circle:

$$QP = r, \quad QO^2 = IO^2 = R^2-2Rr, \quad l \propto a;$$

$$\therefore \quad ar_1^2+br_2^2+cr_3^2 = 2\Delta\,(r+2R).$$

(C) The Nine-point circle:

$$QP = \tfrac{1}{2}R; \quad QO^2 = \tfrac{1}{4}OH^2 = \tfrac{1}{4}R^2-2R\cos A\cos B\cos C.$$

Since the normal coordinates of the nine-point centre are $\cos(B-C)\ldots$, the barycentric coordinates are $\sin A\cos(B-C)\ldots$, so that $l \propto \sin 2B+\sin 2C$.

$$\therefore \quad \Sigma\,(\sin 2B+\sin 2C)\;r_1^2 = 4\Delta\,(1+2\cos A\cos B\cos C).$$

CHAPTER III.

PEDAL AND ANTIPEDAL TRIANGLES.

33. From the point P *within* the circle ABC draw perpendiculars Pd, Pe, Pf to the sides of ABC, and produce AP, BP, CP to meet the circle again in L, M, N, denoting the angles of LMN, as before, by λ, μ, ν.

In the cyclic quadrilateral $PdCe$,

$$Pde = PCA \text{ or } NCA = NLA,$$
$$Pdf = PBA \text{ or } MBA = MLA;$$
$$\therefore \quad d = L = \lambda; \quad \text{so} \quad e = \mu, \quad f = \nu,$$

and the pedal triangle of P is directly similar to LMN.

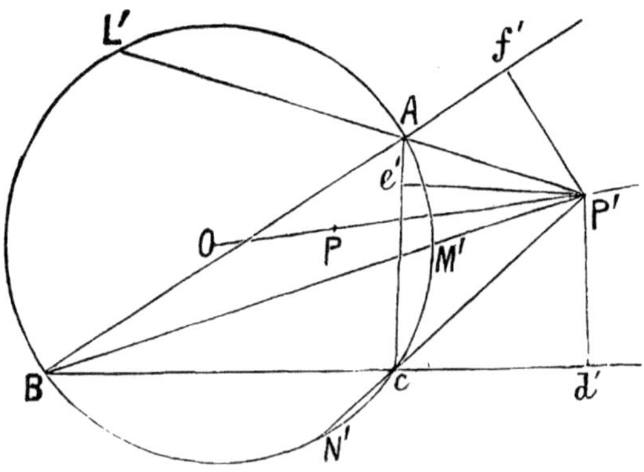

FIG. 11.

34. If P' be the inverse point of P, produce $P'A$, $P'B$, $P'C$ to meet the circle in L', M', N'.

Then $$L' = L = \lambda, \ldots \qquad (21)$$

In the cyclic quadrilateral $P'd'Bf'$,

$$\text{angle } P'd'f' = P'Bf' \text{ or } MBA$$
$$= M'L'A.$$

So $\quad P'd'e' = P'Ce' = N'L'A$ (cyclic quadrilateral $AL'N'C$);

$\therefore \quad d' = P'd'e' - P'd'f' = L' = \lambda, \ldots .$

Hence the pedal triangle of P' is directly similar to $L'M'N'$, which is inversely similar to LMN and def.

It follows that the pedal triangles def, $d'e'f'$ of two inverse points P, P' are *inversely similar.*

35. The sides ef, fd, de of the pedal triangle def, and therefore the sides of LMN, are as $ar_1 : br_2 : cr_3 \quad (AP = r_1)$.

For, since AP is a diameter of the circle $PeAf$,

$$ef = AP \sin A.$$

$$\begin{aligned} \therefore \quad \sin\lambda : \sin\mu : \sin\nu &= ef : fd : de \\ &= MN : NL : LM \\ &= ar_1 : br_2 : cr_3 \\ &= ar_1' : br_2' : cr_3'. \end{aligned}$$

To prove that the area Δ_1 of def

$$= \frac{\Pi\Delta}{4R^2} = \frac{2\Delta^2}{Q}.$$

We have $\quad 2\Delta_1 = de.df.\sin\lambda = r_2 r_3 \sin B \sin C \sin\lambda$

and $\quad 2\Delta = 4R^2 \sin A \sin B \sin C;$

$$\therefore \quad \frac{\Delta_1}{\Delta} = \tfrac{1}{4}\frac{r_2 r_3}{R^2}\frac{\sin\lambda}{\sin A}.$$

But $$r_3 = PM \sin A/\sin\lambda; \qquad (18)$$

$$\therefore \quad \frac{\Delta_1}{\Delta} = \tfrac{1}{4}\,\Pi/R^2 = 2\Delta/Q,$$

since $$\Pi = \frac{2R\,abc}{Q}. \qquad (19)$$

36. All the elements of the pedal triangle def may now be obtained in terms of λ, μ, ν.

For example, to find the circumradius ρ_1

$$2\rho_1^2 \sin\lambda \sin\mu \sin\nu = \Delta_1 = \frac{2\Delta^2}{Q}.$$

Similarly, if Δ_1' be the area of $d'e'f'$,

$$\Delta_1' = \frac{\Pi'\Delta}{4R^2} = \frac{2\Delta^2}{Q'}. \qquad (22)$$

Hence $$\frac{1}{\Delta_1} - \frac{1}{\Delta_1'} = \frac{Q - Q'}{2\Delta^2} = \frac{8\Delta}{2\Delta^2} = \frac{4}{\Delta}.$$

Since $\Pi = R^2 - OP^2$, it follows that the pedal triangles of all points on a circle whose centre is O have the same area.

37. In section (25) the axis $\omega_1\omega_2\omega_3$ is

$$r_1^2\, a\alpha + \ldots = 0 \quad \text{or} \quad \sin^2\lambda\, \frac{a}{\alpha} + \ldots = 0.$$

The pole J of this line is given by

$$\alpha' \propto a(\sin^2\lambda - \sin^2\mu - \sin^2\nu), \ \ldots. \qquad \text{(Salmon, p. 268)}$$

But, since $\sin\lambda$, $\sin\mu$, $\sin\nu$ are proportional to the sides of def,

$$\cos\lambda = \frac{\sin^2\mu + \sin^2\nu - \sin^2\lambda}{2\sin\mu\sin\nu}.$$

Hence, for J, $\qquad \alpha' \propto a\cot\lambda$

or J is $(a\cot\lambda,\ b\cot\mu,\ c\cot\nu)$.

(Dr. J. Schick obtains this result by a different method. See his *Isomorphopolzentrik*, p. 88.)

38. For an example of these formulæ, take the Isodynamic Points U and U'. (26)

Since $\qquad r_1 \propto \dfrac{1}{a},$

$$\therefore \quad ar_1 = br_2 = cr_3;$$

$$\therefore \quad \lambda = \mu = \nu = 60^\circ;$$

so that the pedal triangle def is *equilateral*, and from either U or U' the $\triangle ABC$ is inverted into an equilateral triangle, LMN being similar to def.

$$Q = (a^2+b^2+c^2)\cot 60^\circ + 4\Delta$$
$$= 4\Delta(\cot\omega\cot 60^\circ + 1), \qquad (14)$$

and hence the absolute normal coordinates of U or U' are

$$\alpha = \frac{R}{\cot\omega\cot 60^\circ \pm 1}\,\frac{\sin(A\pm 60^\circ)}{\sin 60^\circ}, \ \ldots. \qquad (19)$$

The powers are $\qquad \dfrac{2R^2}{\cot\omega\cot 60^\circ \pm 1},$

and the areas of the pedal triangles

$$= \frac{1}{2}\,\frac{\Delta}{\cot\omega\cot 60^\circ \pm 1}.$$

39. To determine the power Π of a point P in terms of its barycentric coordinates x, y, z.

Since
$$\frac{a\alpha}{x} = \frac{b\beta}{y} = \frac{c\gamma}{z} = \frac{2\Delta}{x+y+z},$$
$$\therefore \quad \Delta_1 = \tfrac{1}{2}(\beta\gamma \sin A + \dots)$$
$$= \tfrac{1}{2}\frac{4\Delta^2}{(x+y+z)^2}\left\{\frac{y}{b}\frac{z}{c}\frac{a}{2R} + \dots\right\}$$
$$= \frac{\Delta^2}{(x+y+z)^2}\{a^2yz + \dots\}/abcR;$$
$$\therefore \quad \Pi = 4\frac{R^2}{\Delta}\Delta_1 = \frac{a^2yz+\dots}{(x+y+z)^2}.$$

Examples.—

(i) For I, $\qquad x \propto a$;
$$\Pi = \frac{abc}{a+b+c} = 2Rr.$$

(ii) For H, $\qquad x \propto \tan A$;
$$\Pi = \frac{a^2 \tan B \tan C + \dots}{(\tan A + \tan B + \tan C)^2} = 8R^2 \cos A \cos B \cos C.$$

(iii) For K, $\qquad x \propto a^2$;
$$\Pi = \frac{3a^2b^2c^2}{(a^2+b^2+c^2)^2}.$$

(iv) For G, $\qquad x = y = z$;
$$\Pi = \tfrac{1}{9}(a^2+b^2+c^2) = \tfrac{4}{9}\Delta \cot \omega.$$

Let $\qquad AG = m_1 = \tfrac{1}{3}\sqrt{2b^2+2c^2-a^2}, \dots,$

and let $M_1M_2M_3$ be the *Median* Triangle, whose sides are m_1, m_2, m_3.

Producing AG to A_1 and making $GA_1 = GA$, we have the $\triangle A_1BG$ equal in all respects to $M_1M_2M_3$, so that Δ' the area of $M_1M_2M_3 = \frac{1}{3}\Delta$.

To determine M_1,
$$\cot M_1 = \frac{m_2^2+m_3^2-m_1^2}{4\Delta'} = \frac{5a^2-b^2-c^2}{12\Delta},$$
and
$$\cot A = \frac{b^2+c^2-a^2}{4\Delta};$$
$$\therefore \quad \cot M_1 + \cot A = \tfrac{2}{3}\cot \omega.$$

Note that $\qquad \cot M_1 + \cot M_2 + \cot M_3 = \cot \omega,$

so that ABC and $M_1M_2M_3$ have the same Brocard Angle.

The angular coordinates of G are $\pi - M_1, \dots$, so that the angle d of the pedal triangle of $G = X - A = \pi - M_1 - A, \dots$.

The side ef of the pedal triangle $= m_1 \sin A$.

$$\text{Area of pedal triangle} = \frac{\Delta}{4R^2}\Pi = \frac{1}{9}\frac{\Delta^2}{R^2}\cot\omega.$$

40. The Brocard Point Ω or Ω' supplies another interesting illustration of the properties of the pedal triangle.

For Ω, $\qquad A+d = B\Omega C = 180^\circ - C = A+B;\qquad$ (13)

$$\therefore\quad d = B;\quad \text{so}\quad e = C,\ f = A.$$

And if $d'e'f'$ be the pedal triangle of Ω',

$$d' = C,\quad e' = A,\quad f' = B.$$

So that the pedal triangles of the Brocard Points are *similar to* ABC.

To determine ρ, the radius of the pedal circle of Ω,

$$ef = 2\rho \sin d = 2\rho \sin B.$$

But $\qquad ef = r_1 \sin A = 2R\sin\omega\,\frac{b}{a}\sin A;\qquad$ (14)

$$\therefore\quad \rho = R\sin\omega.$$

Hence the linear ratio of the similar triangles def, ABC is $\sin\omega : 1$.

To determine Π for Ω.

We have $\qquad \dfrac{\Delta_1}{\Delta} = \sin^2\omega.$

But $\qquad \Pi = 4\dfrac{R^2}{\Delta}\Delta_1;\qquad$ (35)

$$\therefore\quad R^2 - O\Omega^2 = \Pi = 4R^2\sin^2\omega;$$

$$\therefore\quad O\Omega^2 = R^2(1-4\sin^2\omega).$$

Or Π may be found from the barycentric formula.

For $\qquad l \propto \dfrac{1}{b^2},\quad m \propto \dfrac{1}{c^2},\quad n \propto \dfrac{1}{a^2},\quad$ for Ω;

$\qquad\quad l' \propto \dfrac{1}{c^2},\quad m' \propto \dfrac{1}{a^2},\quad n' \propto \dfrac{1}{b^2},\quad$ for Ω'.

Hence, for Ω

$$\Pi = \frac{a^2b^2c^2}{b^2c^2+c^2a^2+a^2b^2} = \frac{1}{1/a^2+1/b^2+1/c^2} = \Pi';$$

$$\therefore\quad O\Omega = O\Omega'.$$

41. *Equation to a Circle.*—The powers t_1^2, t_2^2, t_3^2 of A, B, C with respect to a circle U being known, to determine the equation to the circle.

Let p, q, r be the perpendiculars from ABC on the Radical Axis of the circles U and ABC, so that $px+qy+rz=0$ is the barycentric equation to the Radical Axis.

Let k be the distance between the centres.

Then, by a well known theorem,

$$\text{Power of } A \text{ for } U = t_1^2 = 2kp.$$

Hence the Radical Axis is $t_1^2x+t_2^2y+t_3^2z=0$.

Let xyz be any point P on U.

Then Π = power of P for ABC

$= 2k \times$ perpendicular from P on Radical Axis

$$= 2k\frac{px+qy+rz}{2\Delta}$$

$$= \frac{t_1^2x+t_2^2y+t_3^2z}{x+y+z}. \qquad (x = a\alpha)$$

But it has been proved that

$$\Pi = \frac{a^2yz+\ldots}{(x+y+z)^2}; \qquad (39)$$

$$\therefore\ a^2yz+b^2zx+c^2xy = (x+y+z)(t_1^2x+t_2^2y+t_3^2z),$$

which is the equation to the circle U.

The former of these two expressions for Π may sometimes be used when the latter is unworkable. For example, to find Π for the Feuerbach Point, whose barycentric coordinates are as $(b-c)^2(s-a)\ldots$.

This point lies on the Nine-point Circle for which

$$t_1^2 = \tfrac{1}{4}(-a^2+b^2+c^2).$$

It also lies on the In-Circle, for which $t_1^2=(s-a)^2$.

Substitute either set of values of t_1^2, t_2^2, t_3^2 in the former expression for Π, and we get

$$\Pi = \frac{1}{4}\frac{\Sigma\, a^3(a-b)(a-c)}{\Sigma\, a\,(a-b)(a-c)};$$

whereas the form of section (39) would give a numerator of the 8th and a denominator of the 6th degree, practically impossible to reduce.

42. *Antipedal Triangles.*—Let S be the point whose angular coordinates are $\pi-\lambda$, $\pi-\mu$, $\pi-\nu$.

Draw QAR, RBP, PCQ perpendicular to SA, SB, SC.

Then PQR is the Antipedal Triangle of S, and its angles are λ, μ, ν.

On SP, SQ, SR as diameters describe the circles $SBPC$, ...
Let AS meet the circle (SP) in A_1.
Then in the triangle BA_1C,

$$\text{angle } BA_1C = BPC = \lambda,$$
$$A_1BC = A_1SC = AQC = \mu,$$
$$A_1CB = A_1SB = ARB = \nu.$$

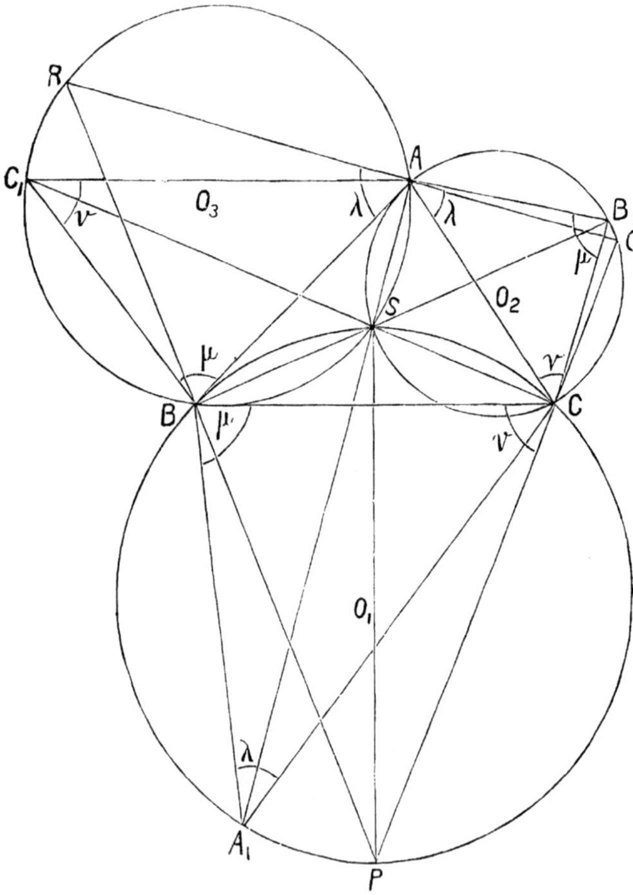

FIG. 12.

Similarly, in the triangle CB_1A,

$$B_1AC = \lambda, \quad CB_1A = \mu, \quad ACB_1 = \nu;$$

and in the triangle AC_1B,

$$C_1AB = \lambda, \quad C_1BA = \mu, \quad AC_1B = \nu;$$

two angles λ being at A, two μ's at B, two ν's at C.

Since SP is a diameter of the circle $SBPC$, the angle SA_1P is right; so that ASA_1, being perpendicular both to A_1P and QAR, is equal to the perpendicular from P on $QR = h_1$.

By Ptolemy's theorem

$$SA_1 . BC = SB . CA_1 + SC . A_1B.$$

But $$BC : CA_1 : A_1B = \sin\lambda : \sin\mu : \sin\nu.$$

$$\therefore \quad SA_1 \sin\lambda = SB \sin\mu + SC \sin\nu.$$

$$\therefore \quad h_1 \sin\lambda = AA_1 \sin\lambda = SA \sin\lambda + SB \sin\mu + SC \sin\nu.$$

43. If X be any point within ABC, the minimum value of $XA \sin\lambda + \ldots$ is $SA \sin\lambda + \ldots$.

Draw XX_1, XX_2, XX_3 perpendiculars to the sides of PQR.

Then $$SA . QR + \ldots = 2\Delta PQR$$
$$= XX_1 . QR + \ldots$$
$$< XA . QR + \ldots;$$
$$\therefore \quad SA \sin\lambda + \ldots < XA \sin\lambda + \ldots.$$

44. *The Isogonic Points.* — The point, within ABC, whose angular coordinates are 120°, 120°, 120°, is called the Inner Isogonic Point, and is denoted by V.

In this case, V is the point for which $r_1 + r_2 + r_3$ is a minimum.

From the preceding article

$$r_1 + r_2 + r_3 = AA_1.$$

And from the triangle ABA_1,

$$AA_1^2 = c^2 + a^2 - 2ac \cos(B + 60°) = 2\Delta(\cot\omega + \sqrt{3})$$

(the triangles A_1BC, B_1CA, C_1AB being equilateral).

$$\therefore \quad (VA + VB + VC)^2 = 2\Delta(\cot\omega + \sqrt{3})$$
$$= \tfrac{1}{2}(a^2 + b^2 + c^2) + 2\Delta\sqrt{3}.$$

45. To determine the area Δ_2 of PQR, the Antipedal Triangle of S, in terms of λ, μ, ν.

Let O_1, O_2, O_3 be the circumcentres of PBC, QCA, RAB, and A', B', C' the mid-points of BC, CA, AB.

Then $$O_1A' = \tfrac{1}{2}a \cot\lambda;$$

$$\therefore \quad \Delta O_1BC = \tfrac{1}{4}a^2 \cot\lambda.$$

Now, O_1 is the mid-point of SP.

$$\therefore \quad \Delta PO_1B = \Delta SO_1B, \quad \Delta PO_1C = SO_1C;$$

$$\therefore \quad \text{area } PBSC = 2\{O_1BC + SBC\}$$
$$= 2\{\tfrac{1}{4}a^2 \cot\lambda + SBC\};$$

$$\therefore \quad \Delta_2 = \text{area of } PQR$$
$$= \tfrac{1}{2}\{a^2 \cot\lambda + b^2 \cot\mu + c^2 \cot\nu + 4\Delta\}$$
$$= \tfrac{1}{2}Q.$$

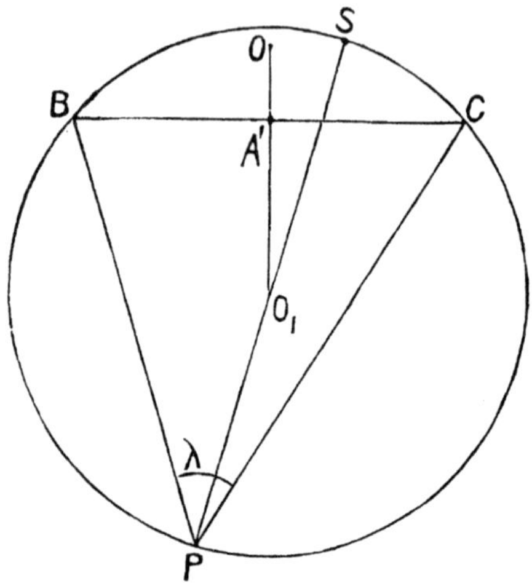

FIG. 13.

If ρ_2 be the circumradius of PQR,

$$4\rho_2^2 \sin\lambda \sin\mu \sin\nu = 2\Delta_2 = Q,$$

and now all the elements of PQR may be determined in terms of λ, μ, ν.

Since $$\Delta_1 = \frac{2\Delta^2}{Q}, \qquad (35)$$

$$\therefore \quad \Delta_1\Delta_2 = \Delta^2.$$

46. To determine (α, β, γ) the normal coordinates of S.

In the circle $SBPC$, $SP = a \operatorname{cosec}\lambda$;

and, since ABC is the pedal triangle of PQR,

$$\text{angle } QSR = P + A = A + \lambda;$$

also $$a\alpha = 2\triangle BSC = SB.SC \sin\lambda.$$

Now $$4\triangle PSR \times \triangle PSQ = SB.SC.PR.PQ$$
$$= \frac{a\alpha}{\sin\lambda}\,\frac{2\triangle QPR}{\sin\lambda};$$
$$= \frac{a\alpha}{\sin^2\lambda}\,Q.$$

Also $4\,\triangle PSR \times \triangle PSQ = SP\,.\,SR \sin(B+\mu)\, SP\,.\,SQ \sin(C+\nu),$

$$= \frac{a^2}{\sin^2\lambda}\,\frac{bc}{\sin\mu\sin\nu}\sin(B+\mu)\sin(C+\nu).$$

Hence $$a = \frac{abc}{Q}\,\frac{\sin(B+\mu)}{\sin\mu}\,\frac{\sin(C+\nu)}{\sin\nu}.$$

47. *Twin Points.*—Just as two points P and P' can be found whose *pedal* triangles are inversely similar, so a companion point S' may be found for S such that the *antipedal* triangles of S, S' are inversely similar.

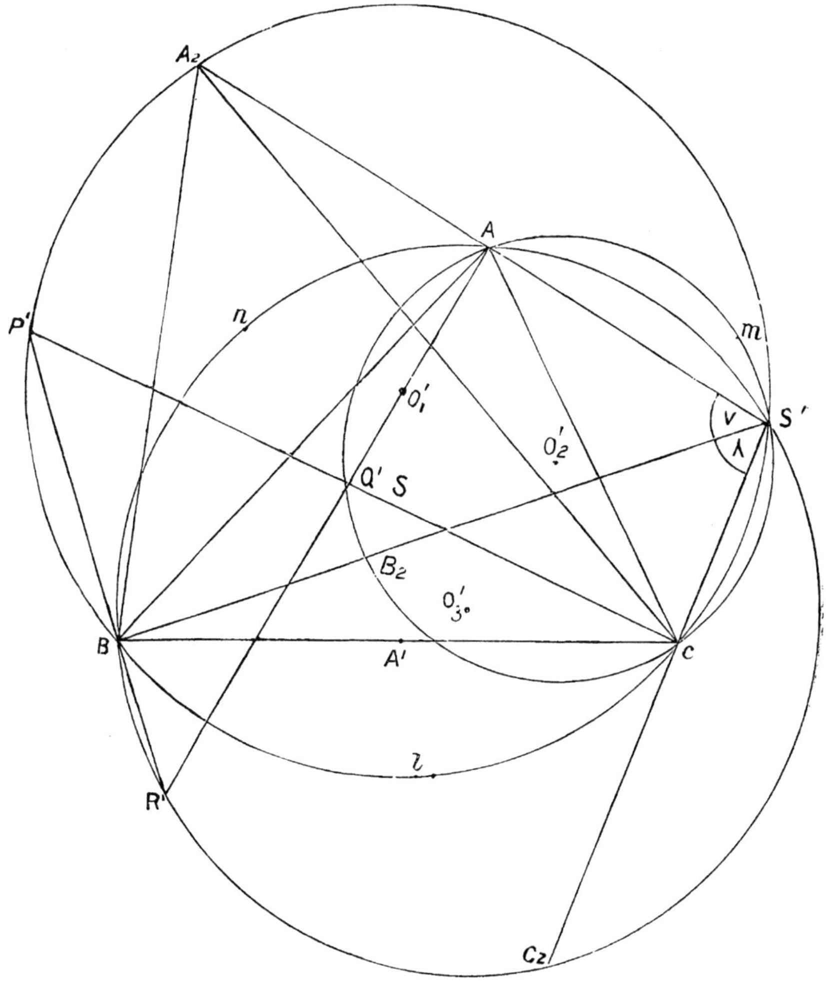

FIG. 14.

Let l, m, n be the images of S in BC, CA, AB, so that the circles BlC, CmA, AnB are the images of the circles BSC, CSA, ASB; and

$$BlC = BSC = 180^\circ - \lambda \dots .$$

Let the circles BlC, CmA cut in S'.

Then $$CS'A = CmA = 180^\circ - \mu = \nu + \lambda.$$

And $$BS'C = 180^\circ - BlC = \lambda.$$

Hence $AS'B = \nu$, so that the third image circle AnB passes through S'.

Draw $AQ'R'$ perpendicular to $S'A$, $BR'P'$ perpendicular to $S'B$, and $CP'Q'$ perpendicular to $S'C$, so that $P'Q'R'$ is the antipedal triangle of S'.

Then $S'CBP'$ is cyclic, since $S'CP' = 90^\circ = S'BP'$; so that P' lies on the circle BlC.

So Q' lies on CmA, and R' on AnB.

Also the angle $Q'P'R'$ or $CP'B = BS'C = \lambda \dots .$

Hence S, S' have inversely similar antipedal triangles with angles λ, μ, ν.

The name "Twin Points" is usually given to S, S'.

Let $S'A$ cut the circle BlC in A_2.

Then $$A_2BC = 180^\circ - CS'A_2 = \mu,$$
$$A_2CB = A_2S'B = \nu;$$
$$\therefore \quad BA_2C = \lambda.$$

So that the triangle A_2BC is the image of A_1BC in BC.

This result gives another method of determining S'.

Take the images A_2 and B_2 of two of the points A_1, B_1, C_1. S' is the point of intersection of AA_2, BB_2.

Note that $$BSC + BS'C = BlC + BS'C$$
$$= 180^\circ;$$
$$ASB + AS'B = AnB + AS'B = 180^\circ,$$
while $$CSA = CmA = CS'A.$$

Hence, if a rectangular hyperbola be drawn through $ABCS$, the point S' is the opposite end of the diameter through S.

The mid-point of SS' is the centre of the rectangular hyperbola, and therefore lies on the Nine-point or Medial Circle.

48. To determine the area (Δ_2') of $P'Q'R'$ in terms of λ, μ, ν.

Let O_1', O_2', O_3' be the centres of the circles $P'BC$, $Q'CA$, $R'AB$;
α', β', γ' the normal coordinates of S';
h_1, h_2, h_3 the perpendiculars from P' on BC, ...

Then, since O_1' is the mid-point of $S'P'$, and lies on OA',

$$h_1 + \alpha' = 2A'O';$$

$$\therefore \quad h_1 - \alpha' = 2\,(A'O_1 - \alpha').$$

$$\therefore \quad \triangle P'BC - S'BC = 2\,\{\triangle O_1'BC - S'BC\}$$
$$= 2\,\{\tfrac{1}{4}a^2 \cot\lambda - S'BC\}.$$

So
$$Q'CA + S'CA = 2\,\{\tfrac{1}{4}b^2 \cot\mu + S'CA\};$$
$$R'AB - S'AB = 2\,\{\tfrac{1}{4}c^2 \cot\nu - S'AB\}.$$

Adding, we have, on the left side,

$$P'BC + R'AB - (ABC - Q'AC)$$
$$= P'BC + R'AB - (Q'BC + Q'AB)$$
$$= (P'BC - Q'BC) + (R'AB - Q'AB)$$
$$= P'Q'R'.$$

Hence
$$\Delta_2' = \tfrac{1}{2}\,\{a^2 \cot\lambda + b^2 \cot\mu + c^2 \cot\nu - 4\Delta\},$$
or
$$\Delta_2' = \tfrac{1}{2}Q'.$$

And
$$\Delta_1' = \frac{2\Delta^2}{Q'}. \qquad (36)$$

Hence
$$\Delta_1'\,\Delta_2' = \Delta^2;$$
and
$$\Delta_2 - \Delta_2' = \tfrac{1}{2}\,\{Q - Q'\} = 4\Delta.$$

The normal coordinates of S' are found to be

$$\alpha' = \frac{abc}{Q'}\,\frac{\sin(B-\mu)}{\sin\mu}\,\frac{\sin(C-\nu)}{\sin\nu} \cdots.$$

CHAPTER IV.

THE MEDIAL TRIANGLE.

49. It is often preferable to take the Mid-Point or Medial Triangle $A'B'C'$, in place of ABC, as triangle of reference.

Let $\alpha\beta\gamma$ and $\alpha'\beta'\gamma'$ be the normal coordinates of a point P referred to the triangles ABC, $A'B'C'$.

$$\text{Then} \qquad \alpha+\alpha' = \tfrac{1}{2}h_1;$$

$$\begin{aligned} \therefore \quad a\alpha+a\alpha' &= \tfrac{1}{2}ah_1 = \Delta \\ &= 4\Delta' \qquad (\Delta' = \text{area of } A'B'C') \\ &= 2\,(a'\alpha'+\ldots) \\ &= a\alpha'+b\beta'+c\gamma'; \\ \therefore \quad a\alpha &= b\beta'+c\gamma'. \end{aligned}$$

$$\text{Hence} \qquad 2\,a\alpha' = -a\alpha+b\beta+c\gamma;$$

or, in barycentric coordinates,

$$x = y'+z',$$

$$2x' = -x+y+z.$$

A line whose ABC equation is

$$l\alpha+m\beta+n\gamma = 0 \quad \text{or} \quad px+qy+rz = 0$$

has for its Medial equation

$$\left(\frac{m}{b}+\frac{n}{c}\right)a\alpha'+\ldots = 0 \quad \text{or} \quad (q+r)x'+(r+p)y'+(p+q)z' = 0.$$

And, conversely, if the Medial equation is

$$l\alpha'+m\beta'+n\gamma' = 0 \quad \text{or} \quad p'x'+q'y'+r'z' = 0,$$

then the ABC equation is

$$\left(-\frac{l}{a}+\frac{m}{b}+\frac{n}{c}\right)a\alpha+\ldots = 0 \quad \text{or} \quad (-p+q+r)x+\ldots = 0.$$

Examples.—

(A) If $H_1H_2H_3$ be the orthic triangle, the ABC equation to H_2H_3 is

$$-\alpha \cos A + \beta \cos B + \gamma \cos C = 0.$$

The Medial equation is

$$a^2 . a\alpha + (a^2 - c^2)\, b\beta + (a^2 - b^2)\, c\gamma = 0.$$

(B) If X, Y, Z are the points of contact of the in-circle, the ABC equation to YZ is

$$-(s-a)\, a\alpha' + \ldots = 0.$$

The Medial equation is

$$a . a\alpha + (a-c)\, b\beta + (a-b)\, c\gamma = 0.$$

50. *The Medial or Nine-point Circle.*—Referred to $A'B'C'$, the equation is

$$\frac{a^2}{x'} + \frac{b^2}{y'} + \frac{c^2}{z'} = 0.$$

Referred to ABC, it is

$$\frac{a^2}{-x+y+z} + \ldots = 0,$$

reducing to the well known form

$$a \cos A\, \alpha^2 + \ldots - a\beta\gamma - \ldots = 0.$$

Or, by the Power method of (41),

$$t_1^2 = \tfrac{1}{2} bc \cos A,$$

so that the circle is

$$2\,(a^2yz + \ldots) = (x+y+z)(bc \cos A\, x + \ldots),$$

reducing to the above form.

The equation of the circumcircle ABC,

$$\frac{a^2}{x} + \frac{b^2}{y} + \frac{c^2}{z} = 0,$$

becomes $$\frac{a^2}{y'+z'} + \frac{b^2}{z'+x'} + \frac{c^2}{x'+y'} = 0,$$

when referred to $A'B'C'$, or

$$a^2x'^2 + \ldots + (a^2+b^2+c^2)(y'z' + z'x' + x'y') = 0.$$

By the Power method,

$$t_1^2 = -A'B . A'C = -\tfrac{1}{4}a^2,$$

$A'B$, $A'C$ being drawn in opposite directions.

$$\therefore \quad (\tfrac{1}{4}a^2y'z' + \ldots) = -(x'+y'+z')(\tfrac{1}{4}a^2x' + \ldots),$$

leading to the above equation.

51. To determine the $A'B'C'$ equation to a circumdiameter TOT', whose direction angles are θ_1, θ_2, θ_3.

If p', q', r' are the perpendiculars on TOT' from A', B', C', its $A'B'C'$ equation is $\qquad p'x+q'y+r'z = 0.$

But a diagram shows that

$$p' = OA' \cos\theta_1 = R\cos A\cos\theta_1.$$

Hence the equation is

$$\cos A\cos\theta_1\, x' + \ldots = 0.$$

Note that, since A' is the mid-point of BC,

$$p' = \tfrac{1}{2}(q+r);$$

and $\qquad p = -p'+q'+r' = R\{-\cos A\cos\theta_1 + \ldots\}.$

52. *A Note on Feuerbach's Theorem.*—Let π be the Feuerbach Point where the in-circle (I) touches the Medial Circle (O').

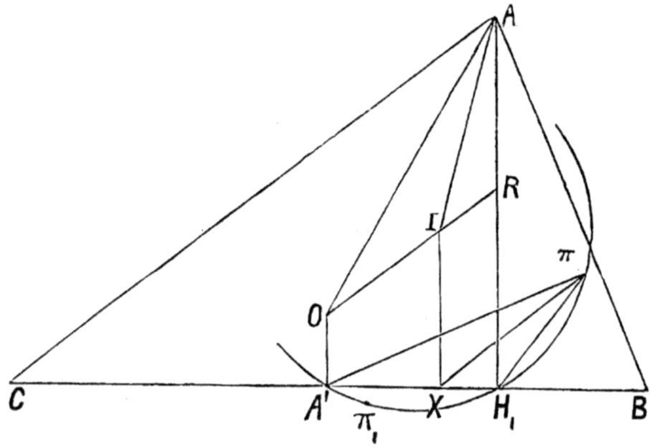

FIG. 15.

Then, since the in-circle also touches $A'H_1$ at X,

$$\therefore\quad \pi X \text{ bisects the angle } A'\pi H_1;$$

$$\therefore\quad A'\pi : \pi H_1 = A'X : XH_1$$

$$= OI : IR = OA : AR,$$

and $\qquad A'\pi H_1 = B - C = OAR.$

Hence the figures $OARI$, $A'\pi H_1 X$ are similar;

$$\therefore\quad A'\pi : A'X = AO : OI, \quad\text{and}\quad A'X = \tfrac{1}{2}(b-c);$$

$$\therefore\quad A'\pi \propto (b-c);$$

and thus the $A'B'C'$ normal coordinates of π are

$$\left(\frac{1}{b-c},\ \frac{1}{c-a},\ \frac{1}{a-b}\right).$$

53. The following is an analytical proof of Feuerbach's Theorem :—

It has been shown that if t_1^2, t_2^2, t_3^2 are the powers of A, B, C for a circle U, then the Radical Axis of the circles U and ABC is
$$t_1^2x+t_2^2y+t_3^2z=0. \qquad (41)$$

Take $A'B'C'$ as triangle of reference.

Then for the in-circle XYZ
$$t_1^2=A'X^2=\tfrac{1}{4}(b-c)^2.$$

Hence the Radical Axis of the circles XYZ, $A'B'C'$ is
$$(b-c)^2x+\ldots=0.$$

But, the barycentric equation to the circle $A'B'C'$ being $\dfrac{a^2}{x}+\ldots=0$, the above is the equation to the tangent to $A'B'C'$ at the point whose barycentric coordinates are
$$\left(\frac{a}{b-c},\ \frac{b}{c-a},\ \frac{c}{a-b}\right)$$
that is, at the Feuerbach Point π.

It follows that the two circles touch at π.

Similarly it may be proved that the Medial Circle touches the ex-circles I_1, I_2, I_3 at the several points π_1, π_2, π_3 whose barycentric coordinates are $\left(\dfrac{a}{b-c},\ -\dfrac{b}{c+a},\ \dfrac{c}{a+b}\right)$, $\left(\dfrac{a}{b+c},\ \dfrac{b}{c-a},\ -\dfrac{c}{a+b}\right)$, and $\left(-\dfrac{a}{b+c},\ \dfrac{b}{c+a},\ \dfrac{c}{a-b}\right)$; the coordinates being obtained by changing successively the signs of a, b, c in the identity
$$a(b-c)+b(c-a)+c(a-b)=0.$$

The tangents to the Medial Circle at π, π_1, ... are therefore
$$(b-c)^2x+(c-a)^2y+(a-b)^2z=0;$$
$$(b-c)^2x+(c+a)^2y+(a+b)^2z=0.$$

Hence, by subtraction, we see that the tangents at π, π_1 intersect on the outer bisector of A', and the tangents at π_2, π_3 on the inner bisector.

54. The following propositions relating to Feuerbach's Theorem are demonstrated in an article by Prof. M. T. Naraniengar, contributed to the *Progress Report of the Indian Mathematical Club*, October, 1908.*

* Published by S. Murthy & Co., 305 Thumbu Chetty Street, Madras.

(A) The straight lines $\pi\pi_1$ and $\pi_2\pi_3$ pass through the feet of the internal and external bisectors of A.

(B) They divide BC harmonically.

(C) L, their point of intersection, has ABC normal coordinates $-\sin(B-C)$, $\sin(C-A)$, $\sin(A-B)$.

If $H_1H_2H_3$ be the orthic triangle, then H_2H_3 passes through L.

(D) L lies on $B'C'$, M on $C'A'$, N on $A'B'$.

(E) AMN, BNL, CLM are straight lines.

(F) AL, BM, CN are perpendicular to the Euler Line $OO'GH$, so that L is the point where the perpendicular from A on OGH cuts $B'C'$.

(G) The triangle LMN is self-polar for the circle $A'B'C'$.

*(H) The polars of A, B, C with respect to the Medial circle pass through L, M, N.

(J) $A'\pi_1$, $B'\pi_2$, $C'\pi_3$ meet two and two on the sides of LMN.

55. A parabola, touching the sides of $A'B'C'$, has TOT' for directrix.

To determine the focus, with $A'B'C'$ as triangle of reference.

It may easily be shown from Salmon's article (*Conics*, p. 274) that, if $(\alpha', \beta', \gamma')$ be the focus, then the directrix is

$$\frac{\cos A}{\alpha'}x + \ldots = 0.$$

This coincides with TOT', whose equation is

$$\cos A\cos\theta_1 . x + \ldots = 0. \qquad (51)$$

Hence the $A'B'C'$ normal coordinates of the focus are

$$\sec\theta_1,\ \sec\theta_2,\ \sec\theta_3\,;$$

a *most important* result.

56. Examples of particular diameters:—

(i) For OI, $\qquad \cos\theta_1 \propto (b-c)$.

The Medial equation to OI is therefore

$$(b-c)\cos A . x + \ldots = 0;$$

and the focus of the parabola is

$$\left(\frac{1}{b-c}\ \frac{1}{c-a}\ \frac{1}{a-b}\right),$$

normal coordinates, or the Feuerbach Point of $A'B'C'$.

* Mr. S. Narayanan has pointed out that these polars also pass through L_1, L_2, L_3, the centres of the Apollonian Circles. (8)

(ii) For OH, $\cos\theta_1 \propto \dfrac{b^2-c^2}{a}$.

The Medial equation to OH is

$$\frac{b^2-c^2}{a}\cos A\,.\,x+\ldots=0;$$

and the focus is $\left(\dfrac{a}{b^2-c^2},\ \dfrac{b}{c^2-a^2},\ \dfrac{c}{a^2-b^2}\right)$,

or the Euler Point of $A'B'C'$.

(iii) For OK, $\cos\theta_1 \propto a(b^2-c^2)$. (11)

The Medial equation to OK is

$$a(b^2-c^2)\cos A\,.\,x+\ldots=0;$$

and the focus is $\left\{\dfrac{1}{a(b^2-c^2)},\ \dfrac{1}{b(c^2-a^2)},\ \dfrac{1}{c(a^2-b^2)}\right\}$,

or the Steiner Point of $A'B'C'$.

57. To determine the focus and directrix of a parabola inscribed in $A'B'C'$ and touching the line whose Medial equation

is
$$\frac{a\alpha}{\lambda}+\frac{b\beta}{\mu}+\frac{c\gamma}{\nu}=0.$$

Since $\dfrac{a\alpha}{\lambda}+\ldots=0$ and $a\alpha+b\beta+c\gamma=0$ (line at ∞),

are tangents to the parabola $\sqrt{(u\alpha)}+\ldots=0$,

$$\therefore\quad \frac{u\lambda}{a}+\frac{v\mu}{b}+\frac{w\nu}{c}=0;$$

$$\frac{u}{a}+\frac{v}{b}+\frac{w}{c}=0;$$

$$\therefore\quad \frac{u}{a}:\frac{v}{b}:\frac{w}{c}=\mu-\nu:\nu-\lambda:\lambda-\mu;$$

therefore the focus is given by

$$\alpha'\propto\frac{a^2}{u}\propto\frac{a}{\mu-\nu};\qquad\text{(Salmon, p. 274)}$$

and the directrix is $(\mu-\nu)\cot A\,.\,x+\ldots=0$.

For example, putting $\lambda\propto a^2$, it is at once seen that the Lemoine Line (8) of $A'B'C'$ touches the parabola inscribed in $A'B'C'$, whose focus is the Euler Point and whose directrix is OH.

58. Let (l, m, n) be the tripolar coordinates of a point T and its inverse T', lying on a fixed circumdiameter whose direction angles are $\theta_1, \theta_2, \theta_3$. (See Fig. 8, p. 16.)

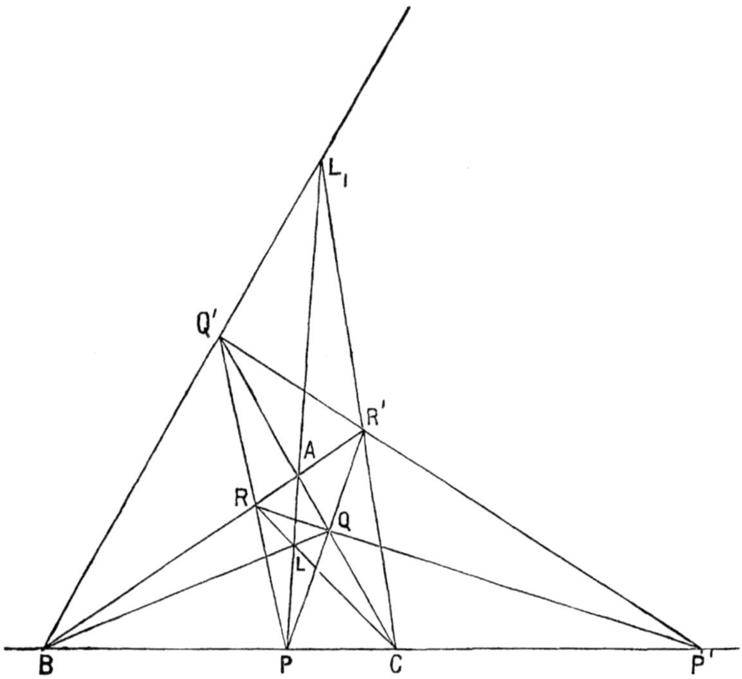

FIG. 16.

Let the internal and external bisectors of the angles BTC, CTA, ATB, or $BT'C$, $CT'A$, $AT'B$, meet the corresponding sides BC, CA, AB in P and P', Q and Q', R and R', so that

$$BP : PC = BP' : P'C = m : n, \ldots.$$

Then AP, BQ, CR are concurrent at a point L, whose barycentric coordinates are $1/l$, $1/m$, $1/n$; while AP, BQ', CR' are concurrent at a point L_1, whose barycentric coordinates are $-1/l$, $1/m$, $1/n$, and so on.

And the triads of points $P'Q'R'$, $P'QR$, $Q'RP$, $R'PQ$ lie on four straight lines forming the system

$$lx \pm my \pm nz = 0$$

(barycentric coordinates being used throughout).

To prove that these four straight lines touch the parabola which is inscribed in the Medial Triangle and has the given circumdiameter for directrix.

The line $$lx+my+nz=0,$$
when referred to the Medial triangle, becomes
$$(m+n)\,x+(n+l)\,y+(l+m)\,z=0. \qquad (49)$$

Now, the barycentric coordinates (x', y', z) of the focus S of the parabola which is inscribed in $A'B'C''$ and touches
$$\frac{x}{\lambda}+\frac{y}{\mu}+\frac{z}{\nu}=0$$
are given by $$x' \propto \frac{a^2}{\mu-\nu}, \ldots. \qquad (57)$$
$$\therefore \quad \left\{\frac{a^2}{m^2-n^2},\ \frac{b^2}{n^2-l^2},\ \frac{c^2}{l^2-m^2}\right\}$$
are the barycentric coordinates of the focus of the parabola which touches the sides of $A'B'C'$, and the line
$$(m+n)\,x+\ldots=0.$$

This focus remaining unchanged by any change of sign in l, m, n, it follows that each of the four lines of the system touches the parabola just determined.

Now draw Tt perpendicular to BC.

Then $$m^2-n^2 \propto TB^2-TC^2 \propto 2a\,.\,A't$$
$$\propto 2a\,.\,OT\,.\cos\theta_1\,;$$
so that the focus is $(a\sec\theta_1,\ b\sec\theta_2,\ c\sec\theta_3)$ and is therefore *fixed*.

The directrix is $$\cos A\cos\theta_1\,.\,x+\ldots=0,$$
and is therefore the circumdiameter TOT'. (51 and 52)

Thus, as T and T' travel along the fixed diameter, the variable system $$lx\pm my\pm nz=0$$
envelops the parabola which touches the sides of $A'B'C'$ and has TOT' for directrix.

Taking the four tangent lines $P'Q'R'$, $P'QR$, $Q'RP$, $R'PQ$ in triads, we see that the four circles PQR, $PQ'R'$, $QR'P'$, $RP'Q'$ pass through the focus, which lies on the Medial circle.

59. With the aid of this parabola we can now deduce some interesting properties of this four-line system.

For these lines, in conjunction with the sides of $A'B'C'$, give a system of seven tangents, forming thirty-five triangles.

The circumcircles of all these triangles pass through the focus S, whose barycentric coordinates are
$$\left\{\frac{a^2}{m^2-n^2}\ldots\right\} \text{ or } \left\{a\sec\theta_1\ldots\right\}.$$

The thirty-five orthocentres lie on the directrix.

$$(m^2-n^2)\cot A.x+\ldots=0;\quad\text{or}\quad \cos A\cos\theta_1.x+\ldots=0.$$

The seven feet of perpendiculars from S on the tangents lie on the vertex tangent, whose equation will prove to be

$$\cot\theta_1.x+\cot\theta_2.y+\cot\theta_3.z=0. \qquad (70)$$

The seven tangents also form thirty-five *quadrilaterals*. The circles described on the diagonals of any one of these quadrilaterals have for their common Radical Axis the four-orthocentre line, which is the fixed directrix TOT'.

In each quadrilateral the mid-point line is perpendicular to the directrix.

60. To determine the *second* point S' in which the circle PQR cuts the Medial Circle.

Let p, q, r be the second points in which the circle PQR cuts the sides of ABC.

Then
$$BP.Bp=BR.Br;$$
$$CQ.Cq=CP.Cp;$$
$$AR.Ar=AQ.Aq.$$

And by Ceva's Theorem
$$BP.CQ.AR=BR.AQ.CP;$$
therefore
$$Bp.Cq.Ar=Br.Aq.Cp,$$
so that Ap, Bq, Cr are concurrent at a point λ_1.

Let
$$Bp:pC=m_1:n_1,\ \ldots,$$
so that $1/l_1$, $1/m_1$, $1/n_1$ are the barycentric coordinates of λ_1.

Now
$$AQ=\frac{l}{n+l}b,\quad AR=\frac{l}{l+m}c,$$
and
$$AQ.Aq=AR.Ar;$$
$$\therefore\quad \frac{a^2}{(m+n)(m_1+n_1)}=\frac{b^2}{(n+l)(n_1+l_1)}=\frac{c^2}{(l+m)(l_1+m_1)},$$
so that
$$l_1 \propto -\frac{a^2}{m+n}+\frac{b^2}{n+l}+\frac{c^2}{l+m}.$$

Hence
$$m_1^2-n_1^2=(m_1+n_1)(m_1-n_1)$$
$$\propto \frac{a^2}{m+n}\left\{\frac{b^2}{n+l}-\frac{c^2}{l+m}\right\}.$$

Hence the barycentric coordinates of S', which are

$$\left\{\frac{a^2}{m_1{}^2-n_1{}^2}\cdots\right\},$$

take the form
$$\left\{\frac{1}{b^2(m+l)-c^2(n+l)}\cdots\right\}.$$

The circle $PQ'R'$ (changing the sign of l) cuts $A'B'C'$ at S and also at the point S_1, whose barycentric coordinates are

$$\left\{\frac{1}{b^2(n-l)-c^2(m-l)}\cdots\right\}.$$

And so for the circles $QR'P'$, $RP'Q'$.

61. $H_1H_2H_3$ being the orthic triangle, to prove that $S'S_1$ and and S_2S_3 intersect on H_2H_3.

As these points are on the circle, $S'S_1$ and S_2S_3 are of the form

$$\frac{a^2}{x_1x_2}x+\ldots=0.$$

The chord $S'S_1$ is

$$a^2\left\{(b^2m-c^2n)^2-(b^2-c^2)^2\right\}x+\ldots=0.$$

The chord S_2S_3 is

$$a^2\left\{(b^2-c^2)^2l^2-(b^2m+c^2n)^2\right\}x+\ldots=0.$$

Adding the two equations, we obtain

$$a^2x+(a^2-c^2)\,y+(a^2-b^2)\,z=0,$$

which is H_2H_3 referred to the Medial triangle. (49)

62. *Examples.*—

(A) Let L coincide with the Gergonne Point, whose barycentric coordinates are $\{1/(s-a)\ldots\}$.

Then P, Q, R are the points of contact of the in-circle, and

$$l\propto s-a : m^2-n^2\propto a(b-c).$$

Hence S is $\{a/(b-c)\ldots\}$ or the Feuerbach Point π; and the directrix is OI, on which lie the orthocentres of PQR, ..., and the points T and T', whose tripolar coordinates are $(s-a)\ldots$.

For S', the second point of intersection of the circle PQR with the Medial circle,

$$x'\propto\frac{1}{b^2(s-b+s-a)-c^2(s-c+s-a)}\propto\frac{a}{b-c}.$$

Hence S and S' coincide, the circle touching the Medial circle at the Feuerbach Point.

The ABC four-line system

$$(s-a)x \pm (s-b)y \pm (s-c)z = 0$$

touches the above parabola.

(B) Let L coincide with I or (abc).

Then AP, ..., are the inner bisectors of A, B, C.

Since $\quad l \propto \dfrac{1}{a} \ldots,\quad$ T and T' are the Isodynamic Points.

Also

$$\frac{a^2}{m^2-n^2} \propto \frac{1}{b^2-c^2};$$

so that S is the Steiner Point, and therefore the directrix is OK, on which lie the orthocentres of PQR, ...

For S', $$x' \propto \frac{1}{b^2\left(\dfrac{1}{b}+\dfrac{1}{a}\right)-c^2\left(\dfrac{1}{c}+\dfrac{1}{a}\right)} \propto \frac{a}{b-c};$$

so that S' is the Feuerbach Point π.

The system $\quad x/a \pm y/b \pm z/c = 0, \quad$ or $\quad \alpha \pm \beta \pm \gamma = 0,$ touches the above parabola.

That in this case the circle PQR passes through the Feuerbach Point was first discovered, I believe, by Mr. J. H. Lawlor, of the College, Maritzburg.

That the Steiner Point is the other point of intersection has been proved independently by Mr. V. Ramaswami Aiyar.

CHAPTER V.

THE SIMSON LINE.

63. From any point T on the circumcircle ABC, draw perpendiculars TX, TY, TZ to the sides BC, CA, AB. Produce TX to meet the circle in t.

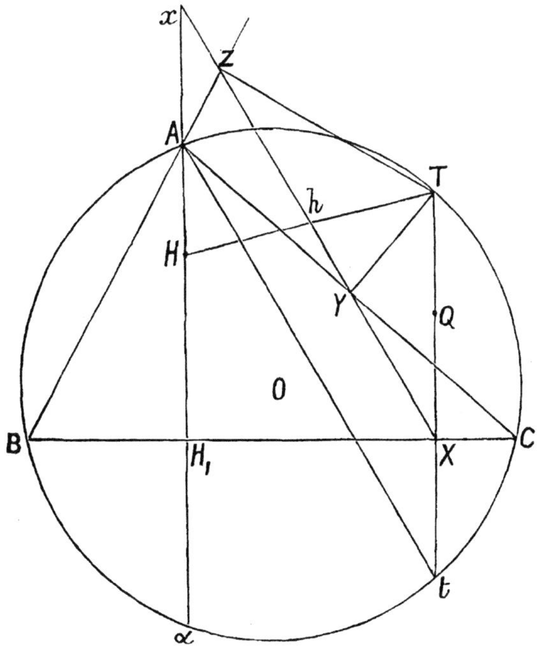

FIG. 17.

Then, since $BZTX$ is cyclic,

$$\text{angle } BZX = BTX \text{ or } BTt = BAt;$$

$$\therefore \quad ZX \text{ is parallel to } At.$$

So XY is parallel to At.

Thus XYZ is a straight line, parallel to At.

It is called the Simson Line of T, and T is called its Pole.

64. To draw a Simson Line in a given direction At.

Draw the chord tT perpendicular to BC, meeting BC in X. A line through X parallel to At is the Simson Line required, T being its pole.

65.* To prove that XYZ bisects TH (H orthocentre).

If Q be the orthocentre of TBC,

$$TQ = 2R\cos A = AH,$$

and
$$QX = Xt = Ax.$$

$$\therefore\quad Hx = TX.$$

$$\therefore\quad HxTX \text{ is a parallelogram.}$$

$$\therefore\quad XYZ \text{ bisects } TH, \text{ say at } h.$$

It follows that h lies on the Nine-point circle.

66. If λ, μ, ν are the direction angles of XYZ, then

$$OAT = OTA = 90^\circ - \tfrac{1}{2}AOT$$
$$= 90^\circ - AtT = 90^\circ - YXT = \lambda.$$

A relation of *fundamental importance.*

67. To determine the absolute normal coordinates of T.

$$TA = 2R.\cos OTA = 2R\cos\lambda.$$

$$\therefore\quad a.2R = TB.TC = 4R^2\cos\mu\cos\nu.$$

$$\therefore\quad a = 2R\cos\mu\cos\nu;$$

so that T is the point ($\sec\lambda$, $\sec\mu$, $\sec\nu$).

Another result of *great use.*

68. To determine the length of the perpendicular from T on its Simson Line.

$$p = TX\sin TXZ = 2R\cos\lambda\cos\mu\cos\nu;$$

or, since
$$TA = 2R\cos\lambda,$$

$$p = \frac{TA.TB.TC}{4R^2}.$$

In (55) the Simson Line of the focus S, being the vertex tangent of the parabola, is parallel to the directrix TOT', so that its direction angles are θ_1, θ_2, θ_3.

Hence the latus rectum = 4 × perp. from S on the Simson Line

$$= \frac{1}{R^2}.SA'.SB'.SC',$$
$$= R\cos\theta_1\cos\theta_2\cos\theta_3.$$

---

* Proof by Mr. N. McArthur. See *Notes* of Edinburgh Mathematical Society, July, 1910.

69. To determine the segments YZ, ZX, XY.

In the circle $AYTZ$, with AT as diameter,

$$YZ = AT \sin A = 2R \cos\lambda \sin A$$
$$= a \cos\lambda\ldots .$$

70. To find the equation to XYZ in terms of λ, μ, ν.

Let u, v, w be the perpendiculars from A, B, C on XYZ.

Then the equation to XYZ is

$$ux + vy + wz = 0.$$

Now
$$u = AY \sin AYZ$$
$$= YZ \frac{\sin AZY}{\sin A} \sin AYZ$$
$$= 2R \cos\lambda \sin A \frac{\sin\nu}{\sin A'} \sin\mu$$
$$= 2R \cos\lambda \sin\mu \sin\nu,$$
$$\propto \cot\lambda.$$

The equation to XYZ is therefore

$$\cot\lambda \,.\, x + \cot\mu \,.\, y + \cot\nu \,.\, z = 0.$$

71. Consider the quadrilateral formed by the sides of ABC and a straight line PQR.

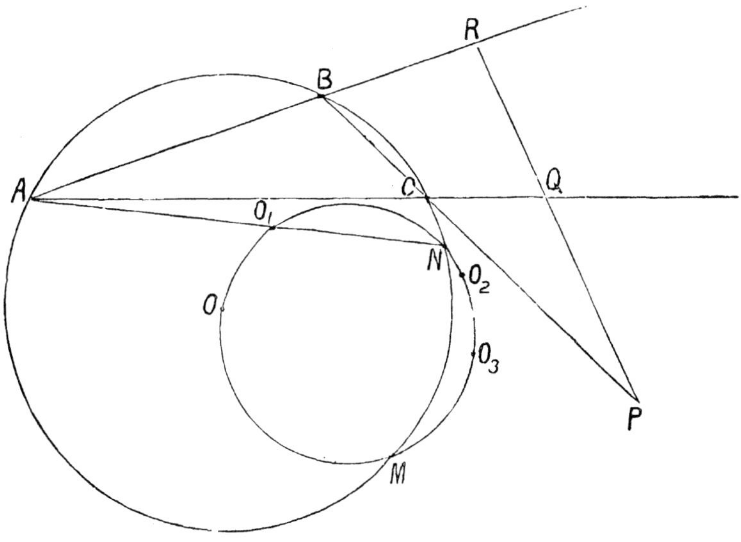

FIG. 18.

The circles ABC, AQR, BRP, CPQ have the common point M, and their centres O, O_1, O_2, O_3 lie on a circle, called the Centre Circle, passing through M.

Let λ, μ, ν be the direction angles of PQR.

If ρ be the circumradius of AQR, the perpendiculars from O_1 on AQ, AR are $\rho \cos R$, $\rho \cos Q$.

So that, if α, β, γ are the normal coordinates of O_1,

$$\beta : \gamma = \cos R : \cos Q = \sec \mu : \sec \nu.$$

Hence AO_1, BO_2, CO_3 meet at a point N whose coordinates are $(\sec \lambda, \sec \mu, \sec \nu)$.

And, since $$\frac{a}{\sec \lambda} + \frac{b}{\sec \mu} + \frac{c}{\sec \nu} = a \cos \lambda + \ldots = 0; \qquad (28)$$

$$\therefore \quad N \text{ lies on the circle } ABC,$$

and is the pole of the Simson Line parallel to PQR.

Again the angle $O_1NO_2 = \pi - ANB$

(BNO_2 being a straight line)

$= \pi - C$.

and since OO_1, OO_2 are perpendicular to MA, MB,

$$\therefore \quad O_1OO_2 = AMB = C;$$

so that N lies on the Centre Circle, and is therefore the second point in which the Centre Circle intersects the circle ABC.

72. T and T' being any points on the circle ABC, it is required to determine S, the point of intersection of their Simson

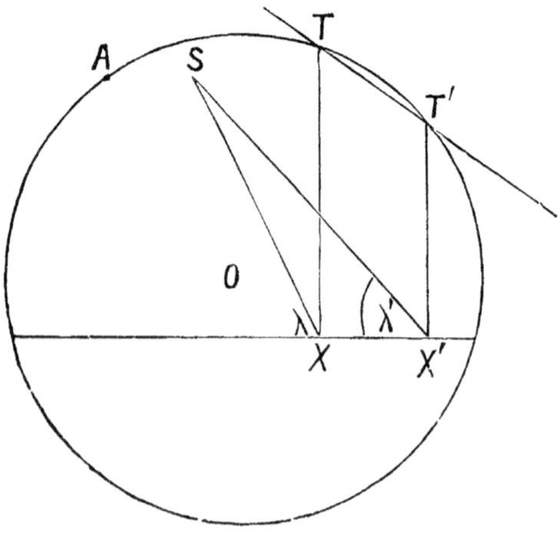

Fig. 19.

Lines. Let $(\lambda\mu\nu)$ $(\lambda'\mu'\nu')$ $(\theta_1\theta_2\theta_3)$ be the direction angles of the Simson Lines of T, T' and of the chord TT'.

Let $(\alpha\beta\gamma)$ be the normal coordinates of S.

Then
$$\alpha = SX \sin\lambda$$
$$= XX' \frac{\sin\lambda \sin\lambda'}{\sin(\lambda-\lambda')}$$
$$= TT' \cos\theta_1 \frac{\sin\lambda \sin\lambda'}{\sin(\lambda-\lambda')}.$$

But
$$OAT = \lambda, \quad OAT' = \lambda', \tag{66}$$

so that
$$TAT' = \lambda-\lambda'.$$

Therefore
$$TT' = 2R \sin(\lambda-\lambda'),$$

and
$$\alpha = 2R \cos\theta_1 \sin\lambda \sin\lambda'.$$

73. To prove that the Simson Lines of T and T', the extremities of a diameter, intersect at right angles on the Medial Circle.

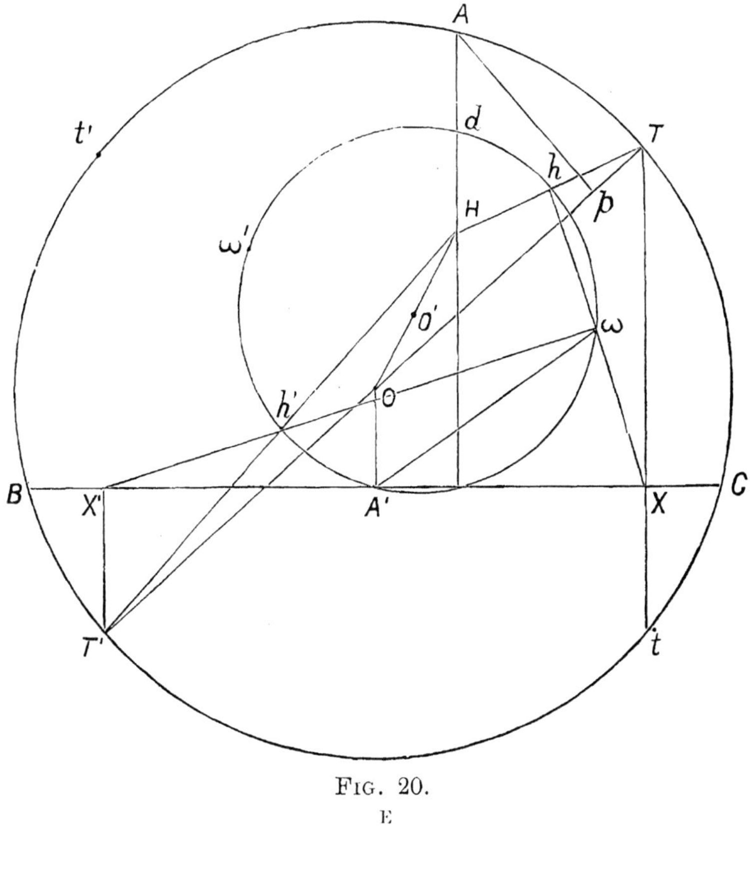

FIG. 20.

Let these Simson Lines Xh, $X'h'$ intersect at ω.

Produce TX, $T'X'$ to meet the circle in t, t', so that $TtT't'$ is a rectangle, and tt' a diameter.

The Simson Lines of T, T' are parallel to At, At', and are therefore at right angles.

Again, since h, h' are mid-points of HT, HT', (65)

$$\therefore \quad hh' = \tfrac{1}{2}TT' = R.$$

But h, h' lie on the Medial Circle; therefore they are the ends of a Medial diameter.

Also $h\omega h' = 90°$, as shown above.

Therefore ω lies on the Medial Circle.

74. If the Simson Lines of *any* two poles T and T' have direction angles $(\lambda\mu\nu)$ and $(\lambda'\mu'\nu')$, the points T, T' are $(\sec\lambda\ldots)$ and $(\sec\lambda'\ldots)$.

Now, the chord through two points $(\alpha_1\beta_1\gamma_1)$ $(\alpha_2\beta_2\gamma_2)$ on the circle is

$$\Sigma\,\frac{a\alpha}{\alpha_1\alpha_2} = 0.$$

Hence, TT' is $\qquad \cos\lambda\cos\lambda'.\,a\alpha+\ldots = 0,$

and the tangent at T is $\quad \cos^2\lambda\,.\,a\alpha+\ldots = 0.$

If TT' be a circumdiameter, then the Simson Line of T' has direction angles $\lambda\pm\dfrac{\pi}{2}\ldots$; so that the absolute normal coordinates of T' are $\quad \alpha' = 2R\sin\mu\sin\nu,\ \ldots$ (67)

and T' is the point $(\operatorname{cosec}\lambda\ldots)$: the equation to the diameter TOT' being $\qquad \cos\lambda\sin\lambda\,.\,a\alpha+\ldots = 0.$

75. Let θ_1, θ_2, θ_3 be the direction angles of a diameter TOT'.

Since $\qquad X\omega X' = 90° \quad$ and $\quad A'X = A'X'$;

$$\therefore \quad \omega \text{ lies on the circle } (XX'),$$

so that $\qquad A'\omega = A'X = R\cos\theta_1.$

At the second point (ω'), where the circle (XX') cuts the Medial Circle, the Simson Lines of t, t' cut at right angles.

The circle (XX') passes through q and r, the feet of the perpendiculars from B, C on TOT'.

For, draw $A'p'$ perpendicular to TOT'; then

$$p'r = p'q = \tfrac{1}{2}qr$$
$$= \tfrac{1}{2}a\cos\theta_1 = R\sin A\cos\theta_1$$

and $\qquad A'p' = R\cos A\cos\theta_1;$ (51)

$$\therefore \quad A'r = A'q = R\cos\theta_1.$$

Note also that $\cot q\omega r = \cot qA'p'$
$= p'A'/p'q = \cot A,$
so that qr, rp, pq subtend angles A, B, $180° - C$ at ω.

Consider the figures $ATOT'$, $\omega XA'X'$.

Since TAT' and $X\omega X'$ are right angles, while O, A' are the mid-points of TOT', $XA'X'$, and

$$\text{the angles } OTA \text{ or } OAT = A'X\omega \text{ or } A'\omega X; \quad (66)$$

it follows that

The figures $ATOT'$, $\omega XA'X'$ are similar,

a fact to be very carefully noted.

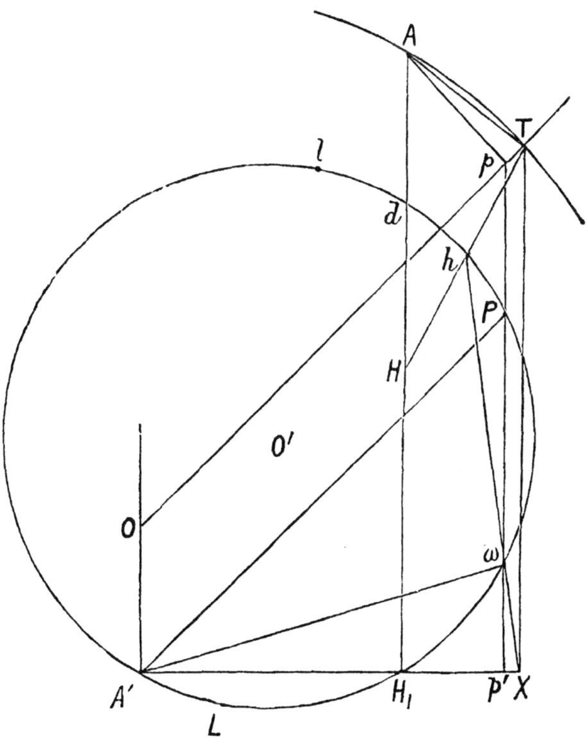

FIG. 21.

Hence angle $\omega A'X = AOT$,
so that ω may at once be found by drawing
angle $XA'\omega = AOT$.

To determine $(\alpha', \beta', \gamma')$ the Medial or $A'B'C'$ coordinates of ω.

Since $\omega' A = R \cos\theta_1 \ldots,$

and $a'.R = \omega B'.\omega C';$

$$\therefore \quad a' = R \cos\theta_2 \cos\theta_3.$$

Hence ω is the point $(\sec\theta_1, \sec\theta_2, \sec\theta_3)$, and its Simson Line (in $A'B'C'$) is parallel to TOT'.

If therefore the chord ωP be drawn perpendicular to BC, then $A'P$, parallel to the Simson Line of ω in $A'B'C'$ (by 63), is also parallel to TOT'.

It will be seen that ω is the focus of the parabola of (52).

76. From the similar figures $ATpO$, $\omega Xp'A'$,

$$A'p' : A'X = Op : OT.$$

Therefore $p'\omega P$ passes through p.

Hence the important proposition—

If Ap, Bq, Cr be drawn perpendicular to a diameter TOT' and pp', qq', rr' be drawn perpendicular to BC, CA, AB, then pp', qq', rr' meet at the point ω on the Medial Circle, which is the point of intersection of the Simson Lines of TOT'; also the (Medial) Simson Line of ω is parallel to TOT'.

77. The envelope of the Simson Line is a Tricusp Hypocycloid.

Since angle $\omega A'H_1 = AOT,$

$$\therefore \quad \omega O'H_1 = 2.AOT. \quad (O' \text{ is Medial Centre})$$

But Medial Radius $= \frac{1}{2}R;$

$$\therefore \quad \text{arc } H_1\omega = \text{arc } AT$$
$$= 2 \text{ arc } dh,$$

by similar figures Hdh, HAT.

Now take arc $A'L = \frac{1}{3}$ arc $A'H_1$,

and draw Medial diameter $LO'l$.

Then, since $A'd$ is also a Medial diameter,

$$\text{arc } A'L = \text{arc } ld;$$
$$\therefore \quad \text{arc } H_1L = 2 \text{ arc } A'L$$
$$= 2 \text{ arc } ld.$$

Also arc $H_1\omega = 2$ arc hd,

as just proved; $\therefore$ arc $L\omega = 2$ arc lh.

Therefore $X\omega h$ touches a Tricusp Hypocycloid, having O' for centre, the Medial Circle for inscribed circle, and $LO'l$ for one axis.

78. If P, P' be any pair of inverse points on the given diameter TOT', then the Simson Lines of T, T' are the axes of similitude, and therefore ω is the double point, for the pair of inversely similar pedal triangles of P, P'. (Neuberg.)

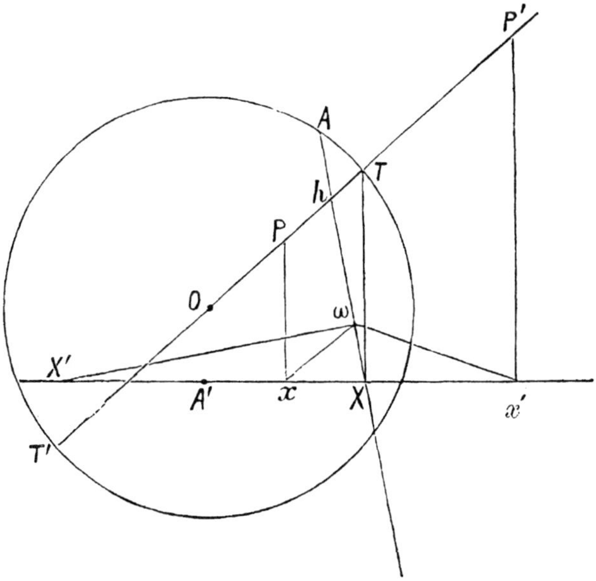

FIG. 22.

In the similar figures $ATOT'$, $\omega XA'X'$ the points P, P' correspond to x, x'.

But AT is known to bisect the angle PAP'; therefore ωX bisects the angle $x\omega x'$.

Also, in the similar figures,

$$\begin{aligned} \omega x : \omega x' &= AP : AP' \\ &= OA : OP' = R : OP' \\ &= \omega y : \omega y' \\ &= \omega z : \omega z'. \end{aligned}$$

Hence ω is the double point, and ωX, $\omega X'$ are the axes of similitude for the inversely similar pedal triangles xyz, $x'y'z'$.

CHAPTER VI.

THE ORTHOPOLE.

79. Let p, q, r be the lengths of the perpendiculars Ap, Bq, Cr from A, B, C on a straight line TT', whose barycentric equation is therefore $px+qy+rz=0.$

Draw pp', qq', rr' perpendicular to BC, CA, AB.

Then shall pp', qq', rr' meet in a point.

A diagram shows that, if θ_1, θ_2, θ_3 are the direction angles of TT', the perpendiculars from A, B, C on pp' are

$$p\sin\theta_1,\quad p\sin\theta_1+c\cos B,\quad p\sin\theta_1-b\cos C,$$

so that pp' is

$$p\sin\theta_1\,x+(p\sin\theta_1+c\cos B)\,y+(p\sin\theta_1-b\cos C)\,z=0;$$

or, noting that $q-r=a\sin\theta_1$,

$$p\,(q-r)(x+y+z)+ac\cos B\,y-ab\cos C\,z=0,$$

with similar forms for qq', rr'.

By addition, all the terms vanish, so that the three lines meet in a point (call it S), which is named by Prof. J. Neuberg the *Orthopole* of TT'.

Solving, we have

$$x\propto q\,(r-p)\,ca\cos B-r\,(p-q)\,ab\cos C+a^2bc\cos B\cos C.$$

80. Let T_1T_2 be any chord of the circle ABC,

T_1X_1, T_2X_2 perpendiculars to BC.

Draw the chord Ap_3t_3 perpendicular to T_1T_2, p_3S and $t_3X_3T_3$ perpendicular to BC, X_3S parallel to At_3, t_3x parallel to BC.

Obviously, SX_3 is the Simson Line of T_3.

We shall prove that SX_1, SX_2 are the Simson Lines of T_1, T_2 respectively.

Since St_3 is a parallelogram,

$$\therefore\quad p_3S=t_3X_3=xX_1.$$

Hence SX_1 is equal and parallel to p_3x.

Again, the line t_3T_1 subtends right angles at p_3 and x, so that $p_3t_3T_1x$ is cyclic.

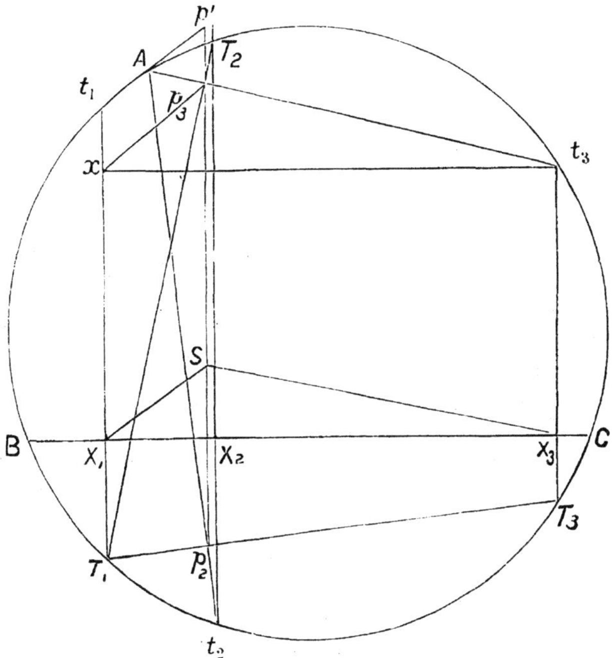

FIG. 23.

$$\therefore \quad \text{angle } p_3xt_1 = p_3t_3T_1 \text{ or } At_3T_1$$
$$= \text{supplement of } At_1x;$$

thus p_3x, and therefore SX_1, is parallel to At_1.

Hence SX_1 is the Simson Line of T_1.

So SX_2 is the Simson Line for T_2, and thus the Simson Lines of $T_1T_2T_3$ are concurrent at S, which lies on the perpendicular from p_3 on BC.

By symmetry, S lies also on the perpendicular from q on CA, and on that from r on AB.

Therefore S is the Orthopole of T_1T_2, so that the Orthopole of a *chord* T_1T_2 is the point of intersection of the Simson Lines of T_1T_2.

It follows at once that S is also the Orthopole of T_1T_3 and T_2T_3.

81. It is clear that each of the points T_1, T_2, T_3 has similar relations to the other two points.

For example.—

(i) SX_3 parallel to At_3 is perpendicular to T_1T_2; so SX_1 is perpendicular to T_2T_3, and SX_2 to T_3T_1.

(ii) The point p_3, the foot of the perpendicular from A on T_1T_2, lies on the perpendicular from S to BC.

Therefore also p_1 and p_2, the feet of the perpendiculars from A on T_2T_3 and T_3T_1, lie on this line.

It follows that $p_1p_2p_3$ is the Simson Line of A for the triangle $T_1T_2T_3$.

Hence the Simson Lines of the poles A, B, C for the triangle $T_1T_2T_3$ pass through S'.

This theorem (ii) is due to Mr. S. Narayanan, B.A.

82. Let $(\lambda_1\mu_1\nu_1)$, $(\lambda_2\mu_2\nu_2)$, $(\lambda_3\mu_3\nu_3)$ be the angles which the Simson Lines of T_1, T_2, T_3 make with the sides of ABC.

The normal coordinates $(\alpha\beta\gamma)$ of S, as the point of intersection of the Simson Lines of T_1 and T_2, are given by

$$x = 2R\cos\theta_1 \sin\lambda_1 \sin\lambda_2, \qquad (72)$$

where θ_1, θ_2, θ_3 are the angles which T_1T_2 makes with the sides of ABC.

But $$\theta_1 = \frac{\pi}{2} - \lambda_3,$$

the Simson Line of T_3 being perpendicular to T_1T_2.

$$\therefore \quad \alpha = 2R\sin\lambda_1 \sin\lambda_2 \sin\lambda_3;$$
$$\beta = 2R\sin\mu_1 \sin\mu_2 \sin\mu_3;$$
$$\gamma = 2R\sin\nu_1 \sin\nu_2 \sin\nu_3.$$

83. To determine the normal coordinates of A with respect to the triangle $T_1T_2T_3$.

These are inversely proportionate to AT_1, AT_2, AT_3.

But $AT_1 = 2R\cos\lambda_1$; $AT_2 = 2R\cos\lambda_2$; $AT_3 = 2R\cos\lambda_3$.

Hence the coordinates are as $\sec\lambda_1$, $\sec\lambda_2$, $\sec\lambda_3$; their absolute values being $2R\cos\lambda_2\cos\lambda_3\ldots$.

84. To prove that S bisects HH'. (H' orthocentre of $T_1T_2T_3$.)

Since $p_1p_2p_3$ is the Simson Line of A for $T_1T_2T_3$, therefore it bisects AH'.

Hence the projection of S on BC is equidistant from the projections of H and H' on BC.

So for CA, AB.

Hence S is the middle point of HH'.

85. To determine Sp_3.

Since t_3A is parallel to SX_3 (Fig. 23),

$$\text{the angle } t_3AB = \text{angle between } SX_3 \text{ and } AB$$
$$= \nu_3.$$

And
$$\text{the angle } t_3BC = t_3AC$$
$$= \text{angle between } SX_3 \text{ and } AC$$
$$= \mu_3.$$

$$\therefore \quad Sp_3 = t_3X_3 = \frac{t_3B\,.\,t_3C}{2R}.$$
$$= 2R\sin t_3AB\sin t_3BC$$
$$= 2R\sin\mu_3\sin\nu_3.$$

86. If T_1T_2 meet the sides of ABC in α, β, γ, to determine the power of S for the circles on $A\alpha$, $B\beta$, $C\gamma$ as diameters.

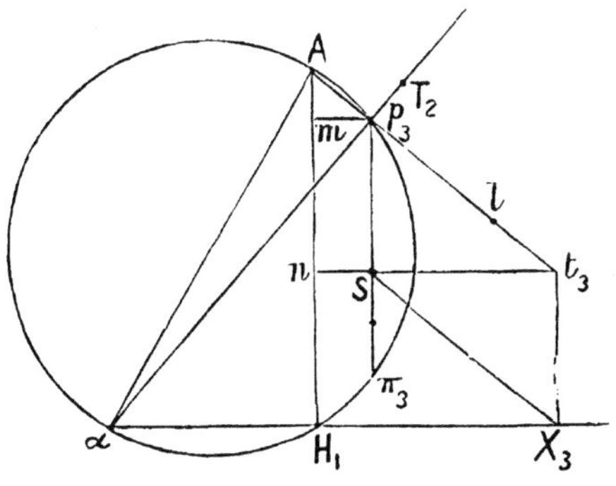

FIG. 24.

The circle $(A\alpha)$ obviously passes through p_3 and H_1, and also through the point π_3, such that $Ap_3\pi_3H_1$ is a symmetrical trapezium.

Then
$$AH_1 - 2Am = p_3\pi_3 = S\pi_3 + Sp_3$$
$$= S\pi_3 + t_3X_3 = S\pi_3 + H_1n;$$
$$\therefore \quad S\pi_3 = AH_1 - H_1n - 2Am$$
$$= An - 2Am$$
$$= (At_3 - 2Ap_3)\sin\lambda_3 \quad (\text{angle } SX_3H_1 = p_3t_3n)$$
$$= p_3l\sin\lambda_3, \qquad \text{where } t_3l = p_3A.$$

Let d_3, ∂_3 be the perpendiculars from O and S on T_1T_2.

Then
$$2d_3 = p_3l.$$

And $\qquad \partial_3 = Sp_3 \sin\lambda_3 = 2R \sin\lambda_3 \sin\mu_3 \sin\nu_3. \qquad (85)$

$\therefore$ Power of S for circle $(A\alpha)$

$$= Sp_3 \,.\, S\pi_3$$
$$= 4Rd_3 \sin\lambda_3 \sin\mu_3 \sin\nu_3$$
$$= 2\partial_3 d_3.$$

The symmetry of the result shows that S has the same power for the three circles $(A\alpha)$, $(B\beta)$, and $(C\gamma)$; therefore it lies on the common Radical Axis of these circles, which is the 4-orthocentre line of the four triangles formed by T_1T_2 and the sides of ABC.

87. *Lemoyne's Theorem.*—TT' is any chord of the circle ABC, the Simson Lines of T, T' intersecting at S.

XYZ is the pedal triangle of any point P on TT'.

To prove that the power of S with regard to the pedal circle XYZ is constant.

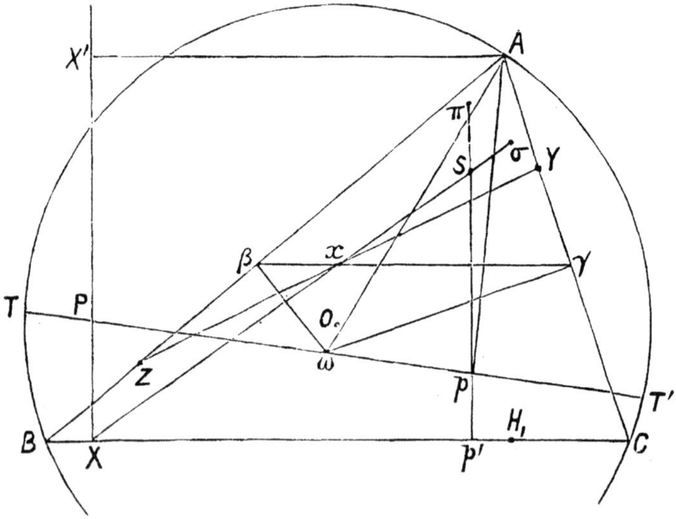

FIG. 25.

Draw Ap, Bq, Cr perpendicular to TT'.

Let TT' cut AO in ω.

Draw $\omega\beta$, $\omega\gamma$ perpendicular to AB, AC, so that $\beta\gamma$ is parallel to BC.

Draw AX' parallel to BC.

Then, since the circle $A\beta\gamma$ with diameter $A\omega$, and the circle AYZ with diameter AP, pass through p, therefore a parabola can be drawn with p for focus, touching AB, AC, $\beta\gamma$, YZ.

Hence the angle $pYC = px\gamma$, since the tangents AC, $\beta\gamma$ cut the tangent YZ in Y and x.

But, from the cyclic quadrilateral $AYpX'$ (diameter AP),

$$pYC = pX'A\,;$$

$$\therefore \quad px\gamma = pX'A,$$

so that pxX' is a straight line.

Hence, as XPX' takes up successive positions $X_1P_1X_1'$, $X_2P_2X_2'$, &c., we shall have

$$x_1x_2 : X_1'X_2' = px : pX' = m, \text{ a constant};$$

$$\therefore \quad x_1x_2 = m\,.\,X_1'X_2' = m\,.\,X_1X_2.$$

It follows that X_1x_1, X_2x_2, &c. are concurrent at some point S'.

But at T and T' the points X, Y, x, Z all lie on the Simson Line of the point.

Therefore S' coincides with S, the Orthopole, or point of intersection of the Simson Lines of T and T'.

Let XS meet the pedal circle XYZ again in σ.

In this circle $$Xx\,.\,x\sigma = Yx\,.\,xZ.$$

And in the circle $AYpZX'$ (diameter AP)

$$X'x\,.\,xp = Yx\,.\,xZ = Xx\,.\,x\sigma.$$

Therefore $X'Xp\sigma$ is cyclic, the centre of the circle lying on the line which bisects XX' at right angles; that is, on the side $B'C'$ of the Medial Triangle $A'B'C'$.

Let π be the image of p in $B'C'$, so that the trapezium $AH_1p\pi$ is symmetrical.

Then, clearly, the circle $X'Xp\sigma$ passes through π; therefore in this circle $$XS\,.\,S\sigma = Sp\,.\,S\pi.$$

But X and σ are points on the pedal circle XYZ;

$$\begin{aligned} \therefore \quad \text{Power of } S \text{ for } XYZ &= XS\,.\,S\sigma \\ &= Sp\,.\,S\pi \\ &= 2\partial d, \qquad (86) \end{aligned}$$

where ∂, d are the perpendiculars from S and O on TT'.

88. *For a circumdiameter T_1T_2 the orthopole S, as the point of intersection of the Simson Lines of T_1T_2, coincides with the point ω (73).*

89. To prove that the pedal circles of all points on the diameter T_1OT_2 pass through the Orthopole ω.

This fact is at once deducible from Lemoyne's Theorem, for now $d = 0$, so that the power of the Orthopole for each pedal circle is zero.

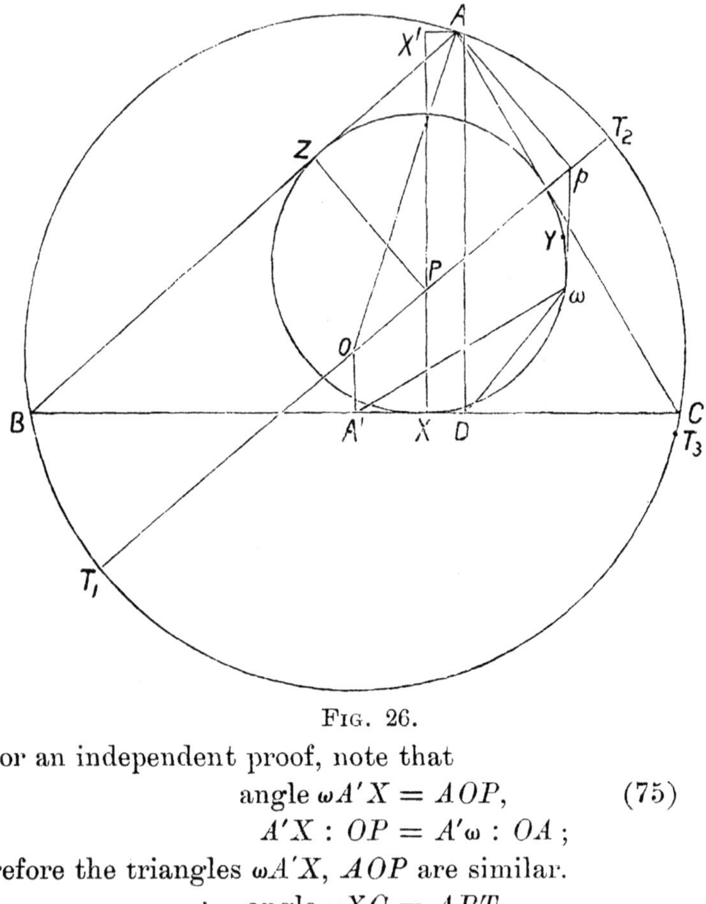

Fig. 26.

For an independent proof, note that

$$\text{angle } \omega A'X = AOP, \qquad (75)$$

and $$A'X : OP = A'\omega : OA;$$

therefore the triangles $\omega A'X$, AOP are similar.

$$\therefore \text{ angle } \omega XC = APT_2.$$

So $$\omega YA = BPT_2,$$

and $$\omega ZA = CPT_2.$$

But $$\omega YA + \omega ZA + Y\omega Z + A = 4 \text{ right angles}$$
$$= BPT_2 + CPT_2 + BPC;$$

$$\therefore \quad Y\omega Z + A = BPC = A + YXZ; \qquad (35)$$

$$\therefore \quad Y\omega Z = YXZ,$$

so that the circle XYZ passes through ω.

Illustrations.—

(*a*) The pedal circles of all points on OI pass through the Feuerbach Point $\left(\frac{1}{b-c}\ \frac{1}{c-a}\ \frac{1}{a-b}\right)$.

(*b*) So for OK and the Steiner Point $\left\{\frac{1}{a\,(b^2-c^2)}\cdots\right\}$.

(*c*) Also for OH and the Euler Point $\left(\frac{a}{b^2-c^2}\cdots\right)$.

Isogonal Conjugate Points.—It is at once seen that the *common* pedal circle of two isogonal conjugates P and P' cuts the Medial Circle in ω and ω', the Orthopoles of OP, OP'. When ω, ω' coincide, or the circles touch, then OPP' is a straight line.

90. To prove that $X\omega$ and YZ intersect on $B'C'$. This follows at once from Lemoyne's Theorem, for there YZ and XS intersect on $\beta\gamma$, which coincides with $B'C'$ when TT' is a diameter.

For an independent proof, draw AX' parallel to BC (Fig. 26). Then, since $AD\omega p$ is a symmetrical trapezium, and DX, AX' are equal and parallel; therefore $X\omega pX'$ is a symmetrical trapezium, and therefore cyclic.

Again, a circle with AP as diameter goes round $PYpAX'Z$, and another circle passes through $X\omega YZ$.

Now let the circles, taken in order, be denoted by U, V, W.

The common chord or Radical Axis of U and V is pX',

that of U and W is ωX,

that of V and W is YZ.

Therefore these chords are concurrent.

But, since the trapezium $X\omega p'X'$ is symmetrical, therefore the chords $X'p$, $X\omega$ intersect on $B'C'$.

$\therefore$ $X\omega$ and YZ intersect on $B'C'$.

NOTE.—Sections 79–80 are based on two articles contributed to *Mathesis* by Prof. J. Neuberg; sections 82–86 and 89 are by the present writer.

CHAPTER VII.

ORTHOGONAL PROJECTION.

*91. AL being a given axis in the plane of the triangle ABC, it is required to determine the shape and size of the orthogonal projection of ABC on a plane X, passing through AL and inclined at an angle θ to the plane ABC.

Let A_1 be the image of A in BC; draw the circle ALA_1 cutting BC again in M, so that $LAM = 90°$.

In AM take a point T, such that $AT/AM = \cos\theta$. Draw BUb, CVc perpendicular to AL, cutting LT in U and V.

Then, since $Ub/Bb = Vc/Cc = AT/AM = \cos\theta$,

AUV represents, in *shape* and *size*, the orthogonal projection of ABC on the plane X.

92. On either side of the common base BC, a series of triangles is described similar to the orthogonal projections of ABC on a series of planes X passing through the common axis AL.

It is required to determine the locus of the vertices of these triangles.

Draw AtS perpendicular to LT, cutting the circle ALA_1 in S'. The triangle SBC will be similar to AUV, and therefore to the projection of ABC on the plane X.

Draw SNS' perpendicular to BC.

Then $$ALT = tAT \text{ or } SAM$$
$$= SLM.$$

And A, t, S, N are right angles.

Therefore the figures $ALtT$, $SLNM$ are similar.

And $$LU : LB = LV : LC = LT : LM.$$

Therefore the figures $ALUVtT$ and $SLBCNM$ are similar.

Therefore the triangle SBC is similar to AUV.

* In writing Sections 91–96 I have drawn freely on Professor J. Neuberg's "Projections et Contre-projections d'un Triangle fixe," with the author's permission.

Hence the vertices of all triangles SBC, described on BC, and having $SBC = AUV$, $SCB = AVU$, $BSC = UAV$, lie on the circle ALA_1.

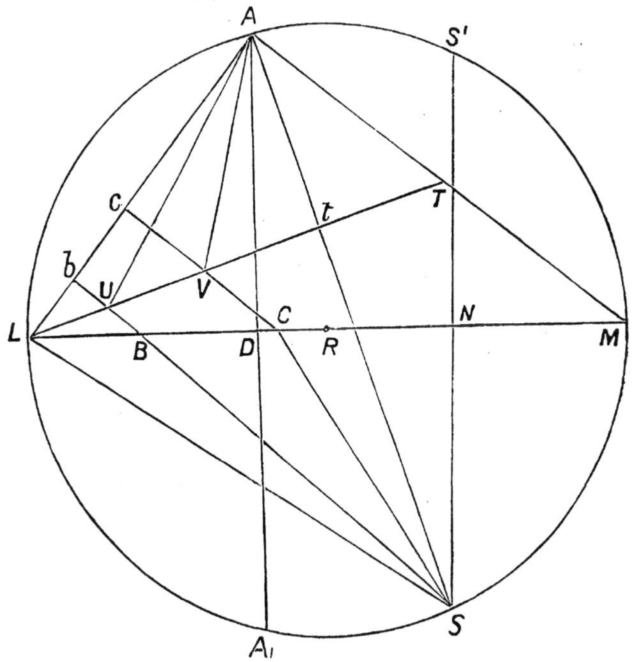

FIG. 27.

The angles of projection range from

$$\theta = 0 \text{ to } \theta = \tfrac{1}{2}\pi.$$

When $\theta = 0$, $\cos\theta = 1$, so that T coincides with M, S with A_1, and S' with A.

As θ increases from 0 to $\frac{1}{2}\pi$, T travels from M to A, S from A_1 to M, and S' from A to M.

Hence the locus of S is the arc AMA_1 on the side of AA' *remote from* the axis AL.

93. When a series of *variable* axes $AL_1, AL_2, \ldots$ are taken, the point A_1, being the image of A in BC, remains unchanged.

Hence each position of the axis AL gives rise to a circle ALA_1M passing through the two fixed points A and A_1, so that this family of circles is *coaxal*.

Let $DA = DA_1 = k$, and let $DR = h$, where R is the centre of the circle ALA_1.

Then, with D as origin and DA as y-axis, the equation of the circle is

$$x^2+y^2-k^2 = 2hx,$$

h being the variable of the coaxal system.

94. We may now deal with the problem: To determine the plane X on which the orthogonal projection of ABC has given angles λ, μ, ν.

Construct the $\triangle SBC$, so that $SBC = \mu$, $SCB = \nu$, and hence $BSC = \lambda$.

Let the circle ASA_1 cut BC in L and M.

Then, if AL and S' are on opposite sides of AA_1, AL is the required axis.

Draw LtT perpendicular to SA.

Then the required inclination θ of the plane X to the plane ABC is given by

$$\cos\theta = \frac{AT}{AM}.$$

95. *Counter-Projection.*—To determine the counter-projection of ABC on a plane through AL, making an angle θ with the plane ABC.

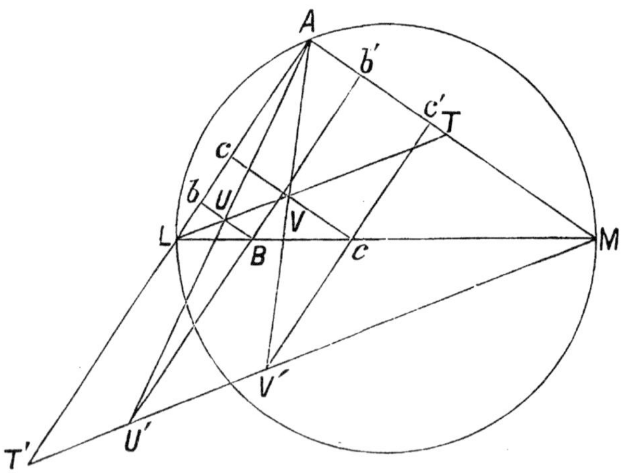

FIG. 28.

Draw MT' parallel to LT; and Bb', Cc' perpendicular equal to AM, meeting MT in U', V'.

Then
$$U'b' : Bb' = T'A : LA$$
$$= MA : TA$$
$$= Bb : Ub;$$
$$\therefore\quad U'b' : Bb = Bb' : Ub.$$

Hence the perpendiculars from U, U' on AL, AM are proportional, and therefore AUU' (and similarly AVV') is a straight line.

If therefore AM be taken as axis, the triangle $AU'V'$, which is the *counter-projection* of ABC, is similar to AUV.

Hence, the variations of the triangle SBC give the shapes of the counter-projections with AM as axis, as well as those of the projections with AL as axis.

It follows that points S on the arc AMA_1 determine the shapes SBC of projections with axis AL, and of counter-projections with axis AM: while points S on the arc ALA_1 determine the shapes of projections with axis AM and counter-projections with axis AL.

96. The triangle ABC being projected on a series of planes making a *constant* angle α with the plane of ABC, and the triangles SBC being drawn, as before, similar to the successive projections, it is required to determine the locus of S.

Draw AL parallel to the line of intersection of the planes, then the original projection is equal and similar to the projection on a parallel plane through AL.

Determine S as before, taking T such that

$$AT/AM = \cos \alpha.$$

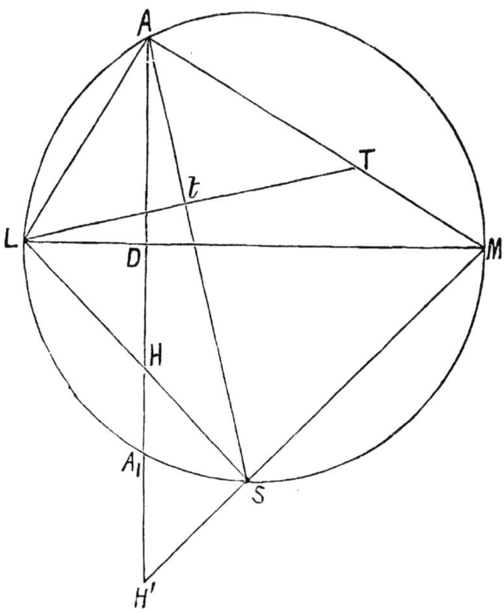

FIG. 29.

Join SL, SM, cutting AA_1 in H and H'.

Then H is the orthocentre of $LH'M$, so that

$$DH.DH' = DL.DM = k^2.$$

Again, $$\cos\alpha = \frac{AT}{AM} = \frac{\tan ALT}{\tan ALM}$$
$$= \frac{\tan SLM}{\tan A_1LM} = \frac{HD}{A_1D};$$
$$\therefore \quad HD = k\cos\alpha = \text{constant},$$
$$H'D = k\sec\alpha = \text{constant},$$
so that H and H' are fixed points.

Hence, since HSH' is a right angle, the point S describes a circle on HH' as diameter.

The equation to the circle HSH' is

$$x^2 + (y - k\cos\alpha)(y - k\sec\alpha) = 0.$$

For another series of planes inclined at a constant angle α', the points H and H' would be changed to H_1 and H_1', where

$$DH_1 = k\cos\alpha', \quad DH_1' = k\sec\alpha',$$
and $$DH_1 . DH_1' = k^2.$$

The series of circles are therefore coaxal, having A and A_1 for limiting points, and therefore cutting orthogonally the former series of circles, which pass through A and A_1.

97. A triangle ABC, with sides a, b, c and area Δ, is projected orthogonally into a triangle LMN, with sides x, y, z and angles λ, μ, ν, its area being Δ', and θ the angle of projection.

It is required to determine the trigonometrical relation between the two triangles.

Let h_1, h_2, h_3 be the heights of A, B, C above the plane LMN.

Then $$h_2 - h_3 = \sqrt{a^2 - x^2};$$
$$\therefore \quad \sqrt{a^2-x^2} \pm \sqrt{b^2-y^2} \pm \sqrt{c^2-z^2} = 0,$$
the signs of the surds depending on the relative heights of A, B, C.

This leads to

$$4y^2z^2 - (y^2+z^2-x^2)^2 + 4b^2c^2 - (b^2+c^2-a^2)^2 = 2\,\Sigma\, x^2(b^2+c^2-a^2).$$

Now $$\cot A = \frac{b^2+c^2-a^2}{4\Delta}, \quad \cot B + \cot C = \frac{a^2}{2\Delta};$$
$$\therefore \quad 16\Delta^2 + 16\Delta'^2 = 16\,\Sigma\,(\cot\mu + \cot\nu)\cot A \,.\, \Delta\Delta';$$
$$\therefore \quad \Sigma\,(\cot\mu + \cot\nu)\cot A \equiv \Sigma\,(\cot B + \cot C)\cot\lambda$$
$$= \Delta'/\Delta + \Delta/\Delta' = \cos\theta + \sec\theta;$$

$$\therefore \quad \Sigma\, x^2 \cot A = 2\Delta' (\cos\theta + \sec\theta),$$

and $$\Sigma\, a^2 \cot\lambda = 2\Delta (\cos\theta + \sec\theta)$$

= sum of areas of projection and counter-projection.

98. *The Equilateral Triangle and the Brocard Angle.*—When ABC is equilateral, we have

$$(x^2+y^2+z^2)\frac{1}{\sqrt{3}} = 2\Delta'(\cos\theta+\sec\theta).$$

But, if ω be the Brocard Angle of LMN,

$$\cot\omega = \frac{x^2+y^2+z^2}{4\Delta'};$$

$$\therefore \quad \cot\omega = \frac{\sqrt{3}}{2}(\cos\theta+\sec\theta).$$

The Brocard Angle therefore depends solely on the angle of projection.

It follows that all coplanar equilateral triangles project into triangles having the same Brocard Angle.

99. *Antipedal Triangles and Projection.*—Let lmn, $l'm'n'$ be the sides, and Δ_2, Δ_2' the areas of the two antipedal triangles, whose angles are λ, μ, ν.

From (45), $$2\Delta_2 = \Sigma\, a^2 \cot\lambda + 4\Delta,$$

and from (48) $$2\Delta_2' = \Sigma\, a^2 \cot\lambda - 4\Delta;$$

$$\therefore \quad \Delta_2 + \Delta_2' = \Sigma\, a^2 \cot\lambda$$

= areas of projection and counter-projection.

Again, $$2\Delta_2 = \Sigma\, a^2 \cot\lambda + 4\Delta$$

$$= 2\Delta(\cos\theta+\sec\theta) + 4\Delta;$$

$$\therefore \quad \Delta_2/\Delta = (\sqrt{\sec\theta}+\sqrt{\cos\theta})^2.$$

So $$\Delta_2'/\Delta = (\sqrt{\sec\theta}-\sqrt{\cos\theta})^2;$$

$$\therefore \quad \sqrt{\Delta_2} - \sqrt{\Delta_2'} = 2\sqrt{\Delta\cos\theta}$$

$$\sqrt{\Delta_2} + \sqrt{\Delta_2'} = 2\sqrt{\Delta\sec\theta}.$$

Now, the triangles Δ_2, Δ_2', with the projection and counter-projection, are all similar, each having angles λ, μ, ν, so that their sides are as the square roots of their areas;

$$\therefore \quad l - l' = 2x, \; \ldots;$$

$$l + l' = 2x, \; \ldots.$$

The theorem, of which the above is a new proof, is due to Lhuillier and Neuberg.

100. The triangle ABC is projected orthogonally on to a plane inclined at an angle θ to the plane of ABC. If S, the line of intersection of the planes, make angles λ, μ, ν with the sides of ABC, it is required to determine the lengths of the sides of the projection in terms of λ, μ, ν, and θ.

Let $$\cos\theta = k.$$

Draw perpendiculars Bb, Cc to the line S, and take
$$bB' = k.bB, \quad cC' = k.cC,$$
so that $B'C'$ is equal to the projection of BC on the plane.

Now, the projection of $B'C'$ on Bb
$$\begin{aligned} &= B'b - C'c \\ &= k\,(Bb - Cc) \\ &= k.a\sin\lambda. \end{aligned}$$

And the projection of $B'C'$ on $S = a\cos\lambda$.

Hence $$\begin{aligned} B'C'^2 = x^2 &= a^2\left\{\cos^2\lambda + k^2\sin^2\lambda\right\} \\ &= a^2\left\{1-(1-k^2)\sin^2\lambda\right\} \\ &= a^2\left\{1-\sin^2\theta\sin^2\lambda\right\}. \end{aligned}$$

The dimensions of the projection depending, of course, only on the *direction*, not on the *position*, of S.

101. To prove that, as the plane of projection revolves round S, the projections of ABC are similar to the pedal triangles of points lying on the circumdiameter TOT', where T is the pole of the Simson Line parallel to S.

Let XYZ and $X'Y'Z'$ be the Simson Lines of T, T'.

Then XYZ, being parallel to S, makes angles λ, μ, ν with the sides of ABC; and therefore
$$\text{angle } OTA = OAT = \lambda.$$

In TOT' take any point P, and let
$$OP = k'.R. \tag{66}$$

Let u, v, w be the sides of the pedal triangle of P.

Then, in the $\triangle OAP$,
$$\begin{aligned} AP^2 &= R^2 + k'^2R^2 - 2k'R^2\cos(\pi-2\lambda) \\ &= R^2(1+k')^2\left\{1-\frac{4k'}{(1+k')^2}\sin^2\lambda\right\}. \end{aligned}$$

So that $$\begin{aligned} 4u^2 &= 4.AP^2\sin^2 A \\ &= a^2(1+k')^2\left\{1-\frac{4k'}{(1+k')^2}\sin^2\lambda\right\}. \end{aligned}$$

But $$x^2 = a^2\left\{1-(1-k^2)\sin^2\lambda\right\}.$$

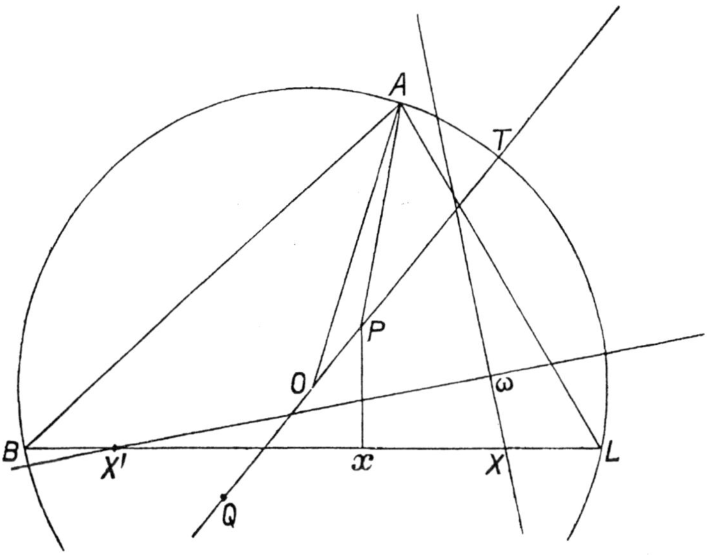

FIG. 30.

Hence $$\frac{x^2}{4u^2} = \frac{1}{(1+k')^2}$$

and $$x : y : z = u : v : w,$$

(so that the projection on the plane θ is similar to the pedal triangle of P,) provided that

$$\frac{4k'}{(1+k')^2} = 1-k^2$$

or $$k = \frac{1-k'}{1+k'}, \quad k' = \frac{1-k}{1+k};$$

so that P, and its inverse point P'', are given by

$$TP/T'P = TP'/T''P' = \cos\theta.$$

The inversely similar pedal triangles of P and P' correspond to the projections of ABC on a pair of planes equally inclined to the plane ABC.

102. If θ_1, θ_2, θ_3 are the angles which TOT' makes with the sides of ABC, the tripolar equation to TOT'' is

$$a\cos\theta_1 . r_1^2 + \ldots = 0. \qquad (27)$$

For P, $$r_1 \sin A = u;$$

$$\therefore \quad \frac{\cos\theta_1}{a} u^2 + \ldots = 0;$$

$$\therefore \quad \frac{\cos\theta_1}{a}x^2 + \ldots = 0,$$

where $$a\cos\theta_1 + \ldots = 0.$$

Hence the sides of the projections are connected by the relation

$$lx^2 + my^2 + nz^2 = 0,$$

with the condition $$la^2 + mb^2 + nc^2 = 0.$$

103. Next, let the plane of projection be inclined at a *constant* angle α to the plane of ABC.

In this case $$k' = \frac{1-k}{1+k} = \frac{1-\cos\alpha}{1+\cos\alpha} = \tan^2\tfrac{1}{2}\alpha;$$

$$\therefore \quad OP = R\tan^2\tfrac{1}{2}\alpha, \quad OP' = R\cot^2\tfrac{1}{2}\alpha.$$

If, therefore, circles be drawn with centre O and radii $\tan^2\frac{1}{2}\alpha \,.\, R$ and $\cot^2\frac{1}{2}\alpha \,.\, R$, any point P on one circle, and its inverse point P' on the other circle, will have their pedal triangles similar to the projections of ABC on a pair of planes drawn through an axis in ABC, and making equal angles α with the plane of ABC.

104. *Counter-Projections.*—To determine the length of the counter-projection of BC on S at the angle θ.

Take $$bB'' = \sec\theta \,.\, bB, \quad cC'' = \sec\theta \,.\, cC,$$

so that $$B''C'' = x'$$

is the required counter-projection of BC.

Now the projection of $B''C''$ on bB

$$= \sec\theta \,.\, a\sin\lambda;$$

$$\therefore \quad B''C''^2 = x'^2 = a^2\{\cos^2\lambda + \sec^2\theta\sin^2\lambda\}$$
$$= a^2\sec^2\theta\{1-(1-k^2)\cos^2\lambda\},$$

showing, as already proved, that the counter-projection of ABC for an axis perpendicular to S is similar to the projection of ABC for the axis S, the angle of projection being θ in each case.

Between O and T' take a point Q, such that

$$OQ = k'R.$$

Then $$OQ^2 = R^2 + k'^2R^2 - 2k'R^2\cos 2\lambda$$
$$= (1+k')^2R^2\left\{1 - \frac{4k'}{(1+k')^2}\cos^2\lambda\right\}.$$

Hence the counter-projections for the axis S are similar to the pedal triangles of points ranging from O to T'.

London:
PRINTED BY C. F. HODGSON & SON,
2 Newton Street, Kingsway, W.C.

Printed in the United States
209213BV00003B/454-456/A

9 781933 998978